THE MIND'S PAST

THE MIND'S PAST

Michael S. Gazzaniga

University of California Press Berkeley Los Angeles London

University of California Press
Berkeley and Los Angeles, California
University of California Press, Ltd.
London, England

Library of Congress Cataloging-in-Publication Data
Gazzaniga, Michael S.
The mind's past / Michael S. Gazzaniga.
p. cm.
Includes bibliographical references and index.
ISBN 0-520-21320-3
(hardcover : alk. paper)
1. Neuropsychology. 2. Brain—Evolution. 3. Memory. 4. Developmental neurobiology. I. Title.
QP360.G392 1998
612.8′2—dc21 97-32505
CIP

Printed in the United States of America
9 8 7 6 5 4 3 2 1
The paper used in this publication meets the minimum requirements of American National Standard for Information Sciences—Permanence of Paper for Printed Library Materials, ANSI Z39.48-1984.

To the memory of

Charlotte Ramsey Smylie,

a great Texan, a great lady,

and a great friend

As long as the brain is a mystery,

the universe will also be a mystery.

SANTIAGO RAMÓN Y CAJAL

CONTENTS

PREFACE

Over a hundred years ago William James lamented, "I wished by treating Psychology like a natural science, to help her to become one." Well, it never occurred. Psychology, which for many was the study of mental life, gave way during the past century to other disciplines. Today the mind sciences are the province of evolutionary biologists, cognitive scientists, neuroscientists, psychophysicists, linguists, computer scientists—you name it. This book is about special truths that these new practitioners of the study of mind have unearthed.

Psychology itself is dead. Or, to put it another way, psychology is in a funny situation. My college, Dartmouth, is constructing a magnificent new building for psychology. Yet its four stories go like this: The basement is all neuroscience. The first floor is devoted to classrooms and administration. The second floor houses social psychology, the third floor, cognitive science, and the fourth, cognitive neuroscience. Why is it called the psychology building?

Traditions are long lasting and hard to give up. The odd thing is that everyone but its practitioners knows about the death of psychology. A dean asked the development office why money could not be raised to reimburse the college for the new psychology building. "Oh, the alumni think it's a dead topic, you know, sort of just counseling. If those guys would call themselves the Department of Brain and Cognitive Science, I could raise $25 million in a week."

The grand questions originally asked by those trained in classical psychology have evolved into matters other scientists can address. My dear friend the late Stanley Schachter of Columbia University told me just before his death that his beloved field of social psychology was not, after all, a cumulative science. Yes, scientists keep asking questions and using the scientific method to answer them, but the answers don't point to a body of knowledge where one result leads to another. It was a strong statement—one that he would be the first to qualify. But he was on to something. The field of psychology is not the field of molecular biology, where new discoveries building on old ones are made every day.

This is not to say that psychological processes and psychological states are uninteresting, even boring, subjects. On the contrary, they are fascinating pieces of the mysterious unknown that many curious minds struggle to understand. How the brain enables mind is *the* question to be answered in the twenty-first century—no doubt about it.

The next question is how to think about this question. That is the business of this little book. I think the message here is significant, one important enough to be held up for examination if it is to take hold.

My view of how the brain works is rooted in an evolutionary perspective that moves from the fact that our mental life reflects the actions of many, perhaps dozens to thousands, of neural devices that are built into our brains at the factory. These devices do crucial things for us, from managing our walking and breathing to helping us with syllogisms. There are all kinds and shapes of neural devices, and they are all clever.

At first it is hard to believe that most of these devices do their jobs before we are aware of their actions. We human beings have a centric view of the world. We think our personal selves are directing the show most of the time. I argue that recent research shows this is not true but simply appears to be true because of a special device in our left brain called the *interpreter*. This one device creates the illusion that we are in charge of our actions, and it does so by interpreting our past—the prior actions of our nervous system. If you want to see how I get there, get from factory-built brain to the serene sense of conscious unity we all possess, you will have to read this mercifully short book.

There are many people to thank, not least of whom is the cleaning lady at the Hotel des Grandes Hommes in Paris. Paris is a wonderful place to launch a book. It feeds you and nourishes you and smiles at you while you struggle away in your small room overlooking a small courtyard. The cleaning lady quickly deduced my assignment and carefully plucked her way around the suitcase of science papers, computers, and espresso cups. Always with a smile, she cheered me on until my wife and children arrived; then, as if handing over a baton in a relay race, she announced to my wife, "Do you know, this is the first time I have seen him smile in days." So many people keep us going.

Of course, scientific guidance has come from many. Steven Pinker once again read and critiqued the whole effort and provided insight upon insight. He is a remarkable scientist and scholar. Michael Posner did the same, and his usual candor and incisive wit straightened me out on several points. To George Wolford, Leo Chalupa, Michael Miller, Ken Britten, Jeffrey Hutsler, Miquel Marin-Padilla, Charlotte Smylie, and many others I offer many thanks. Finally, I benefited from several lectures over the years, not the least of which came from Robert J. Almo, who is not only an orthopedic surgeon, but also an expert on magic.

Alex Meredith, Ph.D., has once again captured the spirit of our work. Many thanks for his brain art in Chapter 5 and chapter openings. They are perfect.

Most important, I thank Howard Boyer at the University of California Press. Now, this is an editor. He not only cleans up the prose, corrects the sentences, clarifies meaning, and encourages one all the way, he is also smart and savvy, sassy and witty. The book would have been much less without him. And, just when you think you are done, along comes the U.C. Press copy editor. She, too, has contributed much to this book; my profound thanks to Sylvia Stein Wright.

Finally, I am reminded of a crack by Pasko Rakic at Yale University. Rakic, one of the world's leading neuroscientists, studies how our cortex develops—a difficult and challenging problem, to say the least. Every sensible scientist stands in nervous awe of the enormity of such task. In reflecting on this, Rakic quipped, "Better not to understand something complex than something simple." My sentiment exactly. Good reading.

MICHAEL S. GAZZANIGA
Sharon, Vermont

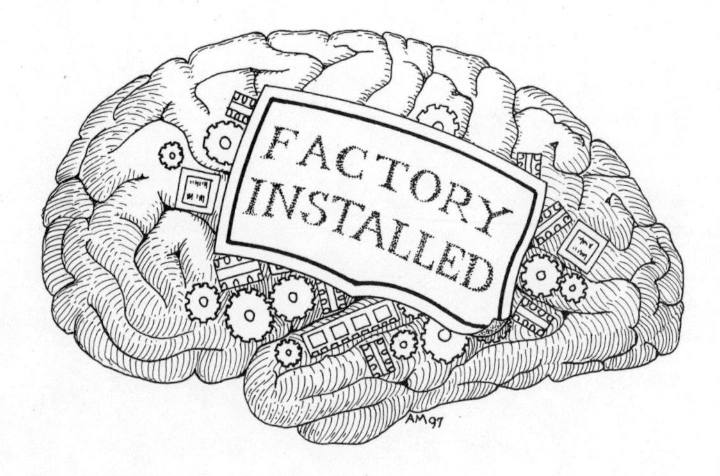
FACTORY
INSTALLED
AM97

1 THE FICTIONAL SELF

There is no life that can be recaptured wholly, as it was. Which is to say that all biography is ultimately fiction. What does that tell you about the nature of life, and does one really want to know?

BERNARD MALAMUD, *Dubin's Lives*

ell, we do know about the fiction of our lives—and we *should* want to know. That's why I have written this book about how our mind and brain accomplish the amazing feat of constructing our past and, in so doing, create the illusion of self, which in turn motivates us to reach beyond our automatic brain.

Reconstruction of events starts with perception and goes all the way up to human reasoning. The mind is the last to know things. After the brain computes an event, the illusory "we" (that is, the mind) becomes aware of it. The brain, particularly the left hemisphere, is built to interpret data the brain has already processed. Yes, there is a special device in the left brain, which I call the *interpreter*, that carries out one more activity upon completion of zillions of automatic brain processes. The interpreter, the last device in the information chain in our brain, reconstructs the brain events and in doing so makes telling errors of perception, memory, and judgment. The clue to how we are built is buried not just in our marvelously robust capacity for these functions, but also in the errors that are frequently made during reconstruction. Biography is fiction. Autobiography is hopelessly inventive.

Over the past thirty years the mind sciences have developed a picture not only of how our brains are built, but also of what they were built to do. The emerging picture is wonderfully clear and pointed. Every newborn is armed with circuits that already compute information enabling the baby to function in the physical universe. The baby does not learn trigonometry, but knows it; does not learn how to distinguish figure from ground, but knows it; does not need to learn, but knows, that when one object with mass hits another, it will move the object.

Even the devices in us that help establish our understanding of social relations may have grown out of perceptual laws delivered to our infant brain. Indeed, the capacity to transmit culture, an act that is only part of the human repertoire of capacities, may grow out of our special capacity to imitate. David and Ann Premack, formerly at the University of Pennsylvania, know a lot about human origins. They have spent much of their careers studying the chimpanzee in the laboratory and have found many instances where the chimp's capacities stop and those of a human infant begin. In their view we uniquely possess many automatic perceptual-motor processes that give rise to the complex array of mental capacities, such as belief and culture.

In considering how much complexity is already built into our brains, I ignore the nature-nurture issue in the traditional sense of how much variance in our intellectual lives is due to our genes and how much to our environment. The issue of whether Billy is smarter than Suzy or vice versa is but frosting on a much bigger cake. I am more concerned with why all humans are different from all chimpanzees, and from any other creature for that matter. Why do we have a theory about our dog or cat, but our cat or dog doesn't have a theory about us? Why don't chimps ever imitate actions or develop a history and a culture, but humans do these things reflexively? That difference is huge. The salient task of this book is to understand how human brains carry out these functions and why no other animal comes close. The brain device we humans possess, which I call the interpreter, allows for special human pursuits. It also creates the impression that our brain works according to "our" instructions, not the other way around.

The way our brains get built and the kinds of circuits that get installed have major consequences. Our brains differ from those of animals. Although our brains are founded on the same building block, the neuron, the organization of these billions of units in our brains gives rise to different capacities. The quantitative differences between Billy and Suzy possibly reflect genetic, intrauterine, and environmental factors. Even IQ differences may represent variations in normal birth trauma; new data suggest that cesarean-delivered infants are brighter. But the qualitative difference in the human brain leads to big discrepancies such as in our capacity for reconstructing past events. This difference deserves our attention. Every normal human, whether a gravedigger or a Nobel laureate, possesses this capacity.

As the Premacks put it, brilliant people like E. O. Wilson at Harvard and Jane Goodall of Tanzania and New Mexico are off-base when it comes to trying to understand the human condition. Wilson claims, "Culture aside from its involvement with language, which is truly unique, differs from animal tradition only in degree." Goodall maintains that, since a chimp cannot talk, it cannot sit down with its peers as humans do and decide what to do tomorrow. The Premacks say, "Animals have neither culture nor history. Furthermore, language is not the only difference between, say, chimpanzees and humans: a human is not a chimpanzee to which language has been added."

My tale weaves its way through what we know about brain development and the simple facts of evolutionary theory as they affect our understanding of the human mind and brain. Even though I constantly call on the insights of biology, I also consider devices in the brain that create a different story for our species. That big, beautiful theory of Charles Darwin, one of the most important scientific theories in the history of the world (and not one word of it was generated with a computer's or calculator's help) leads us to inevitable truths. In attempting to understand what the brain is for, any evolutionary biologist begins with the essential question of why any organ or process does what it does. This approach puts us on a new course in considering how the brain enables mind. Instead of looking for unique physical substrates that support specific functions, we might discover how the brain generates capabilities in informational terms. This is the goal of a serious brain science attempting to understand our psychological selves. Scientists schooled in evolutionary theory keep reminding us of this point. Brain scientists who view the brain as a decision-making device are now gearing their experiments to find answers to the question "What is the brain for?"

The smart-aleck answer to the question is sex. Put more completely, the brain exists to make better decisions about how to enhance reproductive success. Thus, the brain is for helping reproduction and sex. Of course, the body containing it has to survive long enough to have sex. There is little question or disagreement about this. The fun begins with trying to understand how the brain manages this task and where we should look for the answer to the question of the brain's purpose. Most of the scientific observations I report were carried out at the psychological level; this work strongly contributes to the mind sciences, especially when derived from a biological perspective.

. . .

All kinds of things immediately get in our way when we try to think about what the brain does. The human brain, with zillions of capacities and devices for helping us make better decisions about how to enhance our reproductive success, can do many other things along the way. While a computer can be used to compute, which is why it was built, it also makes one hell of a paperweight. The finely tuned human brain can engage primal issues of sexual selection, or it can develop the second law of thermodynamics. Understanding how it does the latter may not inform us of what it normally does and how it does it.

The question "What is the brain for?" is quite different from the question "What can the brain do?" Is this distinction important? So what if brains were built to do X but now serve mostly Y functions, one might argue. It is the Y functions in which scientists are interested. Take reading. Brains were not built to read. Reading is a recent invention of human culture. That is why many people have trouble with the process and why modern brain-imaging studies show that the brain areas involved with reading move around a bit. Our brains have no place dedicated to this new invention, but there is a place that manages breathing. Still, many would say, if the brain accomplishes such a function incidentally to what it was constructed for, so be it.

Most scientists, though, concentrate on the incidental mechanisms, which is a pity. If the evolutionary perspective is simply set aside, the data collected by psychologists and neuroscientists are likely to be grossly misinterpreted. The far-reaching implication of the evolutionary view is that models built to explain psychological and behavioral processes examine only the "noise" of the honed neural system devoted to making decisions about survival. Many psychological models of syntax, for example, assume that a child's ability to master this complex skill simply reflects the manner by which all children come to master the problem of communicating with others. B. F. Skinner, America's and Harvard's most outspoken behaviorist, spent his life promoting his view that such human capacities come about through simple reward contingencies experienced by children. While a proponent of this view would never claim a rat could be taught to talk (since it does not have the innate capacity for that skill), a Skinnerian would maintain that simple reinforcement principles teach an animal or a human everything it is capable of doing.

Nowhere has this Skinnerian view been more prevalent than in explorations of human language. For instance, those who suffered the fifties and sixties heyday of behaviorism and rank empiricism remember being instructed that language is acquired through stimulus and response. Not until Noam Chomsky's pioneering work in linguistics did we realize that language reflects a biological event unique to our species. Many topics that wind up being viewed in evolutionary terms were not illuminated by scientists motivated with that agenda. The irony is that Chomksy, who is anything but a student of evolution, cracked the problem from a totally different perspective—that of the formal analysis of language.

Nonetheless, Chomsky's new view of language as a biologically based universal feature of our brain has taken hold. Steven Pinker, a colleague of Chomsky at MIT, has extended it by successfully arguing that language is an instinct—just like any other adaptation. Syntax is not learned by Skinnerian associative systems; rather, we can communicate through language because all members of our species have an innate capacity to manipulate symbols in a temporal code that maps sounds onto meaning. Although we "learn" different sounds for those meanings, the laws of communication are universal. If an evolutionary perspective were not invoked to interpret the work of linguists, more convoluted psychological theories of learning and development would probably be generated to explain human language, which is in fact an adaptation. Many models have been proposed, but they have little merit or substance.

The debate about the role of evolution and language has produced some strange bedfellows. Stephen Jay Gould sees language as one of his now-famous spandrels, "the tapering triangular spaces formed by intersection of two rounded arches at right angles." Just as these spaces are architectual by-products of mounting a dome onto arches, language, he argues, is simply a by-product of having a big brain. Language came along free with the obvious evolutionary advantage of having a better decision-making device. Oddly, Chomsky seems comfortable with this idea, even though most evolutionary biologists are not.

I think Gould is correct in arguing that there are many spandrels in the mind, but language is not one of them. There are numerous advantages to having language. As the Premacks have pointed out time and again, pedagogy is what our species does best. We are teachers, and we want to teach while sitting by the campfire rather than by being continually present during our offspring's trial-and-error experiences. With language we can communicate both the dangers and the pleasures of the world. Moreover, the advantages of being able to communicate with our nonkin to cooperate in hunting, securing safety, settling disputes, and negotiating a host of daily occurrences are obvious. The appearance of language, slowly but surely gaining in complexity over evolutionary time, can't help but be a huge species event.

Still, I think that many psychological evaluations are superficial. They explain only the noise, or unattended by-products, of a biological system rather than how the system works and what it is capable of doing. They are indeed spandrels.

. . .

In the past decade we have begun to appreciate that the brain is not a big, freewheeling network. It does not make associations based on simple conditional relations and construct from them complex perceptual and cognitive functions. Research in animal psychology, evolutionary psychology, linguistics, and neuroscience has turned to a more fruitful approach to how the brain is structured and how it functions.

This more promising approach is derived from the notion that brains accrue specialized systems (adaptations) through natural selection. These highly specific systems are best understood in relation to their functions. Errors in analysis of their normal function occur when a device proves capable of handling another everyday task and in that capacity appears to have different properties. These proximate properties may be so tangible in our culture that they are accepted as the mechanism involved in the behavior or cognitive function in question.

Modern day illegal drug use, for example, is viewed by some as a deviant behavior produced solely by contemporary social forces. Some discuss at great length reasons for addictions, therapy for addictions, moral implications for drug use, and all the rest. Others simply wonder why we don't use drugs more frequently—if they make us feel good, why not? None of these proximate reasons reveal the underlying forces at work on drug use.

Randolph Nesse at the University of Michigan nails down the reason why drug use occurs and should be dealt with gently. Several adaptations in our brain modulate our emotional states; fear, anxiety, and sadness all help us in our decision making. They are good devices. To highjack these systems with artificial substances is to impair our ability to use cues from the brain systems that manage these emotions and thus to behave in an adaptive way to new challenges. The seriousness of addiction becomes apparent when viewed from the evolutionary perspective. Our built-in systems for cuing good decisions become broken. We are not hearing our normal brain chemical systems advise us on what is good for us. Debates on the morality of addiction and other factors miss this underlying biological issue.

The ebb and flow of neuronal patterns of firing hold the key to how the brain makes its decisions. The physical substrates allow computations to be carried out; but once they are expressed, it is the pattern of the neuronal code that represents the neural code for a function, like seeing a face or a color. Evolutionary theory has generated the notion that we are a collection of adaptations—brain devices that allow us to do specific things. The brain must deal with new challenges in a complex and probably distributed way. Many systems throughout the brain contribute to a single cognitive function. Here's how it all works.

The neural system of any given animal at any given time is in a specific state, but over time the microarchitecture of the neural system changes. If a randomly mutated change—

one which happens when growth dynamics undergo variations—enhances reproductive success, then future generations are likely to inherit the mutation. For example, a rudimentary eye helps an organism to see a little and therefore helps it to navigate the world. If a mutation improves on the rudimentary eye, the organism will see better, behave more efficiently, and survive longer. The genes for that eye become part of the species' hardware. In fact, as Richard Dawkins of Oxford University has recently pointed out, it would not be surprising if all surviving animals possessed some sort of rudimentary eye. It all may have started with a light-sensitive pigment patch that cued the animal as to whether it was night or day. What is known is that all kinds of eyes have evolved, with many species developing unique ways of seeing.

In no way does an organism construct a solution to a problem de novo. Only by chance is a new network generated and additional characteristics and abilities added. Brain mechanisms evolved by random mutation to meet new challenges and perform tasks that enhance reproductive success. This brilliant idea of Darwin is the only explanation for the apparent engineering or complex organic design in the natural world. Trial and error it is, with the "trial" being the random mutation and the "error" being the evidence that the change in the organism is or is not beneficial. This remarkably simple point is still one of the most misunderstood ideas of our time. No matter how eloquently an evolutionist like Richard Dawkins makes the point, people continue to believe it is wrong. "How," they lament, "can a wonderfully complex entity like a human be the product of chance mutations? We must be the result of divine design. We couldn't have happened by chance mutation."

Well, the chance factor is embedded in the idea of natural selection, but at the level of the genome. Chance variations in our genes create potentially better mutations, some of which survive. Over millions of years natural selection works on those chance mutations. Chance mutations and natural selection, working together, produced human beings. But it is completely wrongheaded to say that chance variations in our genome produced us suddenly. It was nibble, nibble, nibble for millions of years.

This ad hoc fashion of building the human, and in particular our brain, unfortunately makes it difficult for neuroscientists to tease out which tasks a system has evolved to accomplish. This is surely why finding localized circuits *wholly* responsible for a perceptual or cognitive capacity is so difficult. A neuropsychologist, the type of scientist who studies the effects of brain lesions on behavior, observes patients with focal lesions—which can result from strokes, tumors, or bullet wounds, or even railroad spikes driven into the skull by explosives. Such a patient may exhibit a specific disorder, such as the inability to detect upright human faces. A neuropsychologist may also study a patient with a large brain lesion that results in an amazingly specific disorder, such as not being able to speak nouns.

. . .

In the profoundly fascinating but young study of how the brain represents adaptive changes, not only within but between species, contemporary knowledge is found wanting.

The human brain is generally regarded as a complex web of adaptations built into the nervous system, even though no one knows how. The neural specificity underlying adaptations probably constitutes a network widely distributed throughout the brain. Since evolutionary changes work in ad hoc ways, often by chance commandeering systems to assist in a chore, the prospects for finding a circuit linked to a task may be very poor.

The powerful theory of natural selection determines how we view the evolved brain and its functions; in accepting this approach we reject traditional behaviorist views of psychology, which posit that our minds are built from simple conditioning and associations. Although the behaviorist view is now out of favor among psychologists, the concept of association networks is currently popular among connectionist theorists. The bastion for these ideas is in La Jolla, California. The intellectual leader of this view is the engaging, clever, and always enthusiastic Terry Sejnowski, a professor sponsored by the Howard Hughes Medical Institute at the Salk Institute by the sea.

Sejnowski and his colleagues believe that genetic specification plays little or no role in the development of our mental devices, and they maintain that neurobiology supports this view. Sejnowski refers to his idea as "neural constructivism," which means that "the representational features of cortex are built progressively from the dynamic interaction between neural growth mechanisms and environmentally derived neural activity. This contrasts markedly with popular selectionist models." While some would argue that constructivism need not necessarily conflict with selectionist views, I think he is right to draw the line at that point. Selectionist models usually refer to how an enviromental stimulus selects out a preexisting capacity that an organism possesses from birth.

Even more boldly, Sejnowski announces that we now know learning guides development; then he quotes a barrage of controversial work. He marries the questionable neurobiology he reviews to the work of Jean Piaget, then suggests children learn domains of knowledge by interacting with the environment. It is not that built-in devices are expressing their capacities.

The La Jolla group draws upon what it calls the "nonstationarity" of the learning device in the brain. Learning transforms the learning device itself, so what has been learned can influence future learning. Sejnowski and his colleagues came to this idea in part because of the difficulties large networks have making guesses about how to organize themselves to solve a problem. They say tiny networks solve small problems and then gradually build up through trial and error.

As David Premack pointed out in his inimitable fashion, the La Jolla group's view of the work amounts to an evolutionist perspective. They want trial and error working ontogenetically, which is to say developmentally; evolutionists have it occurring phylogenetically. The difference is one of time scale, in addition to possible mechanism disparities. But there are deeper problems with connectionist theory. At the level of brain science, the cellular organization of cortical regions can be detected before birth. It is hard to explain to a person who holds a constructivist view that the basic structure of the language cortex is in place before a baby is born. The baby isn't exactly chatting away about Michael Jackson in utero.

Premack points out another problem with the way the La Jolla group thinks about domain-specific knowledge: the fancy way people have come to talk about the fact that what one needs to know about learning language is different from what one needs to know about learning causality. Each has its own domain. The constructivist view of the brain is that it has a common mechanism that solves the structure of all problems. This is aptly dubbed the *problem space*. When the common mechanism confronts language issues, it winds up building the brain one way; when it confronts the problem of detecting faces, it builds it another way—and so on. This sort of assertion leaves us breathless because if we know anything, it is that any old part of the brain can't learn any old thing. Yet Sejnowski and his colleagues strongly believe that the environment plays a major role in structuring the brain and that our experience directly reflects reality. As they say, "This interaction . . . is sufficient to determine the mature representational properties of cortex with no need for domain-specific predispositions somehow embedded *a priori* in the recipient cortex. As a consequence, this makes the relations between environmental changes—whether natural or cultural—and brain structure a direct one." But as Premack says, "When we consider the problems humans are designed to solve, we are struck not by their similarities but by their differences. . . . In the case of language, structure includes phonemes on the one hand, and forms classes, noun vs. verb, on the other. . . . Structure concerns physical relations such as containment, support, collison, and the like in intuitive physics."

The ethological literature contains many examples to counter Sejnowski's claims concerning how our brains are built. Premack reviews the work of Richard Sayfarh on the vervet monkey:

> The vervet's problem domain is predator, the categories of which are: raptor, leopard and snake, to which it produces three different calls. Does the immature vervet figure out the structure of this domain, that is does it learn the categories? No it has the categories: what it learns is how to fine tune the membership of the categories.
>
> For example, a young vervet can mistakenly produce the raptor call to hawks (which resemble the true predator), produce the snake call to inappropriate snakes, and the leopard call to inappropriate ground animals. It corrects these errors, learning to confine the call to the correct member of each category, and to respond more quickly. However, even when the vervet produces its first calls, it does not make between-category errors, e.g., issue the snake call to a bird, etc. Hence, vervets do not "figure out the structure of the problem space." They come with the structure.

. . .

Although many people want to believe that things work the way the La Jolla group claims, biology is not so obliging. The evolutionary engine of natural selection builds beautifully crafted devices in weird ways, creating ever more perfect machines from multitudinous random events.

As a result nature tends to use any trick that comes along to make things work. As George Miller at Princeton University put it when explaining why the cognitive revolution in psychology took place, "During the 1950s it became increasingly clear that behavior is simply the evidence, not the subject matter of psychology." Association theory, behavioral theory, and connectionist theory are inadequate. I am sure we'll find principles that describe our mental activity; that's the goal of the mind sciences. But I am also sure that most of them will be instantiated into complex and possibly bizarre neural devices we are born with, just as we are born with antibodies to ward off other challenges from the environment.

When I minored in immunology in graduate school, the conventional wisdom was that the body can make antibodies to any antigen. This was part of a prevalent view that many things in biology can be formed by instruction—that the organism incorporates information from the environment into its function. This is exactly what the neural constructivists are saying today. The idea in immunology was that globulin molecules will form around any antigen and generate an immune response. But by the mid-1960s the revolution in biology had demonstrated that nothing is farther from the truth. It had become clear that organisms, from mice to humans, are born with all the antibodies they will ever have. Instead of the body responding to instruction from the environment, the invading antigen selects a group of antibodies already in the body. These antibodies are amplified and produce the classical immune response. The immune system's complexity is thus built into the body. So is the mind's complexity.

Niels Jerne, the brilliant immunologist and Nobel laureate, played a primary role in alerting neuroscientists to the value of biological mechanisms such as selection theory in understanding how the brain enables mind. Jerne pointed out that, despite the long-held belief that biological processes are subject to instruction, every time we figure out a biological process, it is selection, not instruction, that is at work. The finches in the Galapagos Islands, for instance, suffered a drought in 1977. Those with large beaks could make use of a more varied food supply. And so, after a generation or two, smaller-beaked individuals died off, and large beaks became a dominant feature among the finch population. It wasn't that the small-beaked finches had not learned to grow larger beaks to eat the new food supply; rather, the genetic characteristic of large-beakedness was rapidly selected for. The same process occurs in the well-known mutation of microorganisms into resistant strains when antibiotics are used inappropriately. Although instruction may appear to be at work, as if the environment were "telling" organisms to change, selection is in fact calling the shots.

Jerne hypothesized that the nervous system is constructed by a similar process. Perhaps much of what passes for learning (i.e., the body receiving instruction from the environment) is more akin to selection. Jerne suggested that an organism has a cornucopia of genetically determined neural networks for certain kinds of learning. When learning, the system sorts through an array of built-in responses, seeking the appropriate one to use for the environmental challenge.

A more cognitive modification of Jerne's idea would

take into account one of the primary findings of cognitive science over the past forty years. We don't select our ideas preformed, just as we don't select sentences preformed from some inventory, like Tickle-Me Elmo dolls. We form sentences combinatorially, as Chomsky showed the world, with a computational apparatus that recursively combines preexisting elements into bigger and bigger structures. The same is true for our thoughts. The human's devices for meeting new challenges allow for a seemingly endless array of inventive solutions to environmental challenges. But those devices are built in and are brought to bear on new problems.

Jerne's original hypothesis, modified or not, is bold; yet so much of what we now know about the brain, animal behavior, psychological development, evolution, and human neuropsychology is consistent with it. The brain is a collection of systems designed to perform functions that contribute to the primary goal of every brain: to enhance reproductive success. As Paul Rozin, a psychologist from the University of Pennsylvania, noted years ago, just as one can view intelligence as a phenotype and look at the multitude of subprograms that contribute to a skill, one can view human cognition as a phenotype and identify subprograms that make up this feature of brain activity. It is the accumulation of additional circuits that accounts for the unique human experience.

. . .

Funnyman Henny Youngman often said, "Timing is everything." With our brains chock full of marvelous devices, you would think that they do their duties automatically, before we are truly aware of the acts. This is precisely what happens. Not only do automatic mechanisms exist, but the primate brain also prepares cells for decisive action long before we are even thinking about making a decision! These automatic processes sometimes get tricked and create illusions—blatant demonstrations of these automatic devices that operate so efficiently that no one can do anything to stop them. They run their course and we see them in action; as a consequence we have to conclude that they are a big part of us.

Our motor system, which makes operational our brain's decisions about the world, is independent of our conscious perceptions. Too often our perceptions are in error; so it could be disastrous to have our lives depend on them. We would be better off if our brains reacted to real sensory truths, not illusory ones.

If so many processes are automatic, they should function outside of our conscious awareness. But we have come to think that the part of our brain which has grown like Topsy, the cerebral cortex, is reserved for conscious activities. Brain scientists have been wrongheaded about this, too. The cortex is packed with unconscious processes, as are the older parts of our brain.

Imagine that fate has not been good to you, and you suffer a cerebral stroke. It could have been worse, but it does destroy the primary visual system of your brain's right half. You no longer can see anything to the left of your primary visual focus. Reports over the past twenty years indicate that while you may not consciously see in your blind field of vision, your hand or even your mouth might be able to respond to stimuli presented there. Patients who exhibit this condition, which has been dubbed *blindsight*, can respond to such stimuli without being consciously aware of them. To explain this odd finding, researchers proposed that the deep, dark, phylogenetically older parts of the midbrain, not the cortex, were carrying out the task. They hoped that the site of unconscious processing had been discovered and could be examined. The promise was great, but this idea has been questioned and is probably wrong. There is no need to look for the unconscious in the midbrain; it is upstairs in the cortex, right where it belongs.

Ever since Freud introduced his psychodynamic ideas, people have been fascinated with the unconscious. There, in that mysterious domain of our mental life, ideas are strung together, true relations between facts are seen, plans are made. Although Freud never specified which parts of the brain manage the unconscious, the tacit assumption and sometimes explicit claim is that the older and more primitive regions do this work. Consciousness is rooted in the cerebral cortex—the great big blanket that covers our older midbrain and hindbrain. But, so the theory goes, the mysteries of life—what we do outside of conscious awareness—stir in the Cimmerian depths below.

In the collective enthusiasm for this simplistic thinking, we all missed a fundamental point: Ninety-eight percent of what the brain does is outside of conscious awareness. No one would disagree that virtually all our sensorimotor activities are unconsciously planned and executed. As I sit here and type this sentence, I have no idea how my brain directs my fingers to the correct keys on the keyboard. I have no idea how the bird sitting on the outside deck, a glimpse of which I must have caught in my peripheral vision, just snagged my attention while I nonetheless continue to type these words. The same goes for intellectual behaviors. As I sit and write, I am not aware of how the neural messages arise from various parts of my brain and are programmed into something resembling a rational argument. It all just happens.

Surely we are not aware of how much of anything gets done in the realm of our so-called "conscious" lives. As we use one word and suddenly a related word comes into our consciousness with a greater probability than another, do we really think we have such processes under conscious control?

Our mind has an absurdly hard time when it tries to control our automatic brain. Remember the night you woke up at 3 A.M., full of worry about this and that? Such concerns always look black in the middle of the night. Remember how you tried to put them aside and get back to sleep? Remember how bad you were at it?

We all have had our interest sparked by an attractive stranger. A struggle ensues as we try to override the deeply wired brain circuitry provided by evolution to maintain our desire to reproduce. Allaying possible embarrassment, the mind gets around the brain's assertion this time and manages to maintain control. Society does have an effect, through yet other brain representations, and thus we are not completely at the mercy of our brain's reproductive systems. Or at least we like to believe so.

Why do some of us like going to work so much? There goes that brain again. It has circuits that need attention, that want to work on problems. Then comes the weekly lunch, complete with fine wine, delicious food, and stimulating conversation. Mr. Brain, there you go again. I suppose you will want to sleep after lunch, too.

Nowhere is the issue of ourselves and our brain more apparent than when we see how ineffectual the mind is at trying to control the brain. In those terms, the conscious self is like a harried playground monitor, a hapless entity charged with the responsibility of keeping track of multitudinous brain impulses running in all directions at once. And yet the mind is the brain, too. What's going on?

There seems always to be a private narrative taking place inside each of us. It consists partly of the effort to fashion a coherent whole from the thousands of systems we have inherited to cope with challenges. Novelist John Updike muses on this in his book *Self Consciousness*:

> "Consciousness is a disease," Unamuno says. Religion would relieve the symptoms. Religion construed, of course, broadly, not only in the form of the world's barbaric and even atrocious religious orthodoxies but in the form of any private system, be it adoration of Elvis Presley or hatred of nuclear weapons, be it a fetishism of politics or popular culture, that submerges in a transcendent concern the grimly finite facts of our individual human case. How remarkably fertile the religious imagination is, how fervid the appetite for significance; it sets gods to growing on every bush and rock. Astrology, UFOs, resurrections, mental metal-bending, visions in space, and voodoo flourish in the weekly tabloids we buy at the cash register along with our groceries. Falling in love—its mythologization of the beloved and everything that touches her or him is an invented religion, and reli-

> gious also is our persistence, against all the powerful post-Copernican, post-Darwinian evidence that we are insignificant accidents within a vast uncaused churning, in feeling that our life is a story, with a pattern and a moral and an inevitability—that as Emerson said, "a thread runs through all things: all worlds are strung on it, as beads: and men, and events, and life come to us, only because of that thread." That our subjectivity, in other words, dominates, through secret channels, outer reality, and the universe has a personal structure.

Indeed. And what in our brains provides for that thread? What system ties the vast output of our thousands upon thousands of automatic systems into our subjectivity to render a personal story for each of us?

A special system carries out this interpretive synthesis. Located only in the brain's left hemisphere, the interpreter seeks explanations for internal and external events. It is tied to our general capacity to see how contiguous events relate to one another. The interpreter, a built-in specialization in its own right, operates on the activities of other adaptations built into our brain. These adaptations are most likely cortically based, but they work largely outside of conscious awareness, as do most of our mental activities.

The left-hemisphere interpreter was revealed during a simultaneous concept test in which split-brain patients were presented with two pictures. One picture was shown exclusively to the left hemisphere and the other exclusively to the right. The patient was asked to choose from an array of pictures ones that were lateralized to the left and right sides of the brain. In one example, a picture of a chicken claw was flashed to the left hemisphere and a picture of a snow scene to the right hemisphere. Of the array of pictures placed in front of the subject, the obviously correct association was a chicken for the chicken claw and a shovel for the snow scene. One of the patients responded by choosing the shovel with his left hand and the chicken with his right. When asked why he chose these items, his left hemisphere replied, "Oh, that's simple. The chicken claw goes with the chicken, and you need a shovel to clean out the chicken shed." In this case the left brain, observing the left hand's response, interpreted that response in a context consistent with its sphere of knowledge—one that does not include information about the snow scene.

What is amazing here is that the left hemisphere is perfectly capable of saying something like, "Look, I have no idea why I picked the shovel—I had my brain split, don't you remember? You probably presented something to the half of my brain that can't talk; this happens to me all the time. You know I can't tell you why I picked the shovel. Quit asking me this stupid question." But it doesn't say this. The left brain weaves its story in order to convince itself and you that it is in full control.

The interpreter influences other mental capacities, such as our ability to accurately recall past events. We are poor at doing that, and it is the interpreter's fault. We know this because of neuropsychologists' research on the problem. My favorite comes from studies of the two half brains of split-brain patients. The memory's accuracy is influenced by which hemisphere is used. Only the left brain has an interpreter, so the left hemisphere has a predilection to interpret events that affect the accuracy of memory. The interpreterless right hemisphere does not. Consider the following.

When pictures that represent common events—getting up in the morning or making cookies—were shown to a split-brain patient who was later asked to identify whether pictures in another series had appeared in the first, both hemispheres were equally accurate in recognizing the previously viewed pictures and rejecting unseen ones. But when the subject was shown related pictures that had not been shown, only the right brain performed well. The left hemisphere incorrectly recalled more of these pictures, presumably because they fit into the schema it had constructed regarding the event. This finding is consistent with the idea of a left-hemisphere interpreter that constructs theories to assimilate perceived information into a comprehensible whole. In so doing, however, the elaborative processing has a deleterious effect on the accuracy of reconstructing the past.

What is so adaptive about having what amounts to a spin doctor in the left brain? Isn't telling the truth always best? In fact, most of us are lousy liars. We become anxious, guilt ridden, and sweaty. As Lillian Hellman observed, guilt is a good thing; it keeps us on the straight and narrow. Still, the interpreter is working on a different level. It is really trying to keep our personal story together. To do that, we have to learn to lie to ourselves.

Robert Trivers pointed this out years ago, as I reviewed in my last book, *Nature's Mind*. In order to convince someone else of the truth of our story we have to convince ourself. We need something that expands the actual facts of our experience into an ongoing narrative, the self-image we have been building in our mind for years. The spin doctoring that goes on keeps us believing we are good people, that we are in control and mean to do good. It is probably the most amazing mechanism the human being possesses.

Come along with me as I weave my way through the mind and the brain. Let me show you why I think our interpreter is reconstructing the automatic activities of our brain. Let me tell you about how our brain is built, how it makes mistakes, how it gets things done for us, and how we put on it a spin that makes it all seem like we are personally in charge. I don't know if I have it all right, but I'm confident I don't have it all wrong!

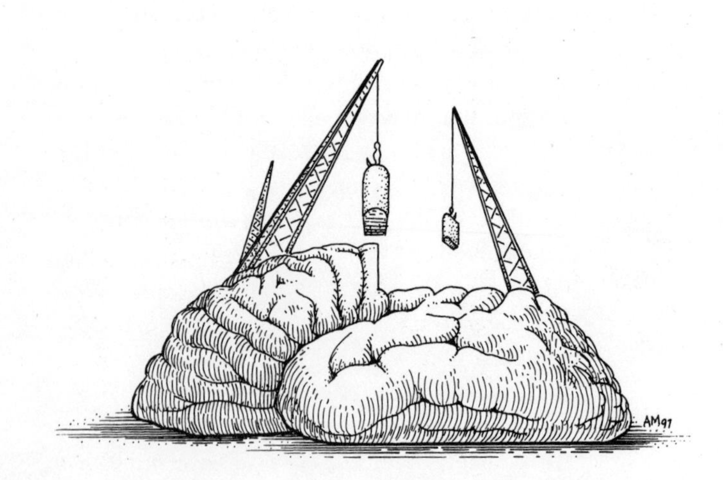
AM97

2 BRAIN CONSTRUCTION

The one fact that I would cry from every housetop is this: the Good Life is waiting for us—here and now! . . . At this very moment we have the necessary techniques, both material and psychological, to create a full and satisfying life for everyone.

B. F. SKINNER, *Walden Two*

ne sunny spring day in 1997 Bill and Hillary Clinton held a White House conference on babies and brains. Psychologists and neuroscientists, eager to please the president and first lady, painted a pretty picture of brain development implying that the more you read to your babies, the better off they will be. The assertion seemed so reasonable that the *New York Times* published an editorial saying neuroscience had informed us that the brain needs crafting during development and reading is the way to do it. Such happy news. All brains can be built to be all things. Spending time with your children and reading to them are good ideas because brain scientists and the *New York Times* said so.

This kind of casual reasoning drives serious scientists to distraction. Neuroscience has plenty to say about brain development, and developmental psychologists have performed numerous clever experiments on the nature of the young mind. Yet nowhere in all their results is there evidence that reading to the young builds better brains. Many scientists have considered to what extent the brain is built along genetically determined lines and how much its neural connections can be changed by experience. Most students of brain development believe it is an ordered process that creates a remarkably intricate device that is mainly wired by the time of birth. Yet scientists who study the adult brain on molecular and cellular levels tend to believe in plasticity, that the brain rewires itself through experience. After one conference, which I arranged and then heard the titans of the two approaches go on and on for three days, I wondered whether they were talking about the same organ. Their radically different views were confusing, to say the least.

Brian Smith, a computer scientist at Indiana University, reviewed the issues of mental development and accused fellow scientists of talking at one level of analysis and then, in midsentence, switching to another level in a way that left listeners breathless. These science commentators have no interest in or capacity for filling in the intervening steps. Smith likened this trend in science to a tunnel diode that at one moment is in state X and then at another is in state Y. That's OK for diodes but not for human reasoning. It is not admissible to be speaking about brain development and suddenly draw conclusions about the wisdom of reading to a child. Perhaps those arguing for plasticity, while armed with intriguing phenomena, read into analyses more than is there. There are thousands of intervening steps that must be understood.

Consider the poor *New York Times* editorial writer. He listens to scientists extrapolating their work and then passes their projections along, unexamined, to the general public. But even he should be able to ask, "So the brain responds differently to a parent reading and to a radio playing Pearl Jam?" That question would make neuroscientists run for cover. No research has differentiated between brain development patterns from two such diverse experiences. The claim that the brain benefits specifically from reading is an egregious example of political correctness.

The *New York Times* editorial also drew upon a *Times* story wherein Dr. William Staso, a school psychologist in Orcutt, California, who wrote a book called *What Stimulation Your Baby Needs to Become Smart* reveals his thoughts about the brain. In a boxed highlight for the article Dr. Staso has become an "expert in neurological development." Staso and the writer cite some goofy study purporting to show that verbal input to babies is important for success in life because it disclosed that children of professional parents hear at home an average of twenty-one hundred words an hour, the children of working-class parents hear twelve hundred words an hour, and welfare parents deliver only six hundred words an hour. The offspring of professionals do better in life, so early reading must be the answer! The study is almost hilariously bad—political correctness drives the science.

There are numerous differences between the upper and lower classes, including genes. Being innately articulate is one way to enter and stay in the upper classes; conversely, if you're always tongue-tied, you'll probably be downwardly mobile. Parents who talk a lot may have kids who do well because they share the same genes for talking and reading. The crucial test would be to study adopted kids only or parents and the children they gave up for adoption. My bet is the correlation would go completely with the genes, not with the upbringing. That happens with everything else looked at this way.

The *Times* augmented President Clinton's argument by quoting the usually talented psychologist Patricia Kuhl of the University of Washington: "We now know that neural connections are formed very early in life and that the infant's brain is literally waiting for experiences to determine how connections are made. We didn't realize until very recently how early this process begins. For example infants have learned the sounds of their native language by the age of six months." Interestingly, the *Times* calls her a neuroscientist, not a psychologist, and then winds up with the social policy recommendation that the state intervene in young children's lives earlier than it does now. People learn things, to be sure, and the brain does this learning. To say the brain is waiting for information is to say nothing of interest. To note that babies learn the sounds of their native language says nothing about the wisdom of reading to children to enhance the structure of the brain.

All this excitement harks back to earlier observations about rats and how much more developed their brains appeared to be if they grew up in an enriched environment than in an isolated cage. Put them in with pals and toys and lots of action and, bingo, fat cortex. So read to your kids and. . . . Well, every social worker knows the home environment of the inner-city child does not lack stimulation. There is usually an extended family, many children,

TV constantly on, and a host of everyday problems being dealt with. If anything, the quiet suburban household might be the perfect place to carry out a sensory deprivation experiment.

But the need to satisfy certain political norms affects scientific thinking. When science is mixed with social policy issues, scientists can and frequently do lose their grip. They may present their results in a light pleasing to the political system they are beholden to. It is a problem all public-minded scientists struggle with every day. It is hard to be totally objective and truthful, especially in the presence of the president.

Of course, there is nothing wrong with reading to babies. It surely must be a good thing. It builds culture and interpersonal relations; it supplies information. Those are good things. But as John Bruer, president of the James S. McDonnell Foundation, has said, the bridge between neuroscience and education is a bridge too far. Don't read to babies because you think it builds better brains. Read to them because you want to be with them and to begin their education. Reading is a good thing! Our culture no longer seems able to say, "Read. It's good for you." Everything has gone over to the health idiom: "Read. It's good for your brain." Whatever happened to the idea that reading is pleasurable in and of itself?

In any brain-mind account it is crucial to know how the brain is built and how vulnerable it is to change. This is too important an issue to be glib about in the East Room of the White House. Can key events going on during development be tweaked to create better brains-minds? Can things happen that create worse brains-minds? Does brain development take place independently of subtle environmental influences?

These issues play out in one of two intellectual contexts. Traditional psychologists, as we found in Chapter 1, always believed in the powerful roles of associative learning in normal adult development. British empiricists from John Locke on initially believed that ideas are linked: Young minds are built from simple associations, and each developed idea hooks up with others. Then in the early part of this century America's experimental psychologists proceeded for over fifty years to write the rules of simple learning.

The contrasting view of how to think about development comes from the natural sciences and ultimately from the ideas of evolutionary biology. Over the years animal behaviorists have revealed the ingenious ways animals develop adaptations or capacities to cope with challenges they face in their own niches. From an ethological viewpoint creatures are born with complexity built into them—complexity established by the mechanisms of natural selection.

A long struggle between these two views seems likely. Each of us brings to the objective data in front of us an implicit view of how the world works. Personal and formal political or social leanings color our rational processes in such a way that we can render almost any observation either critical or meaningless, as we see fit. This has to stop if we are going to build a clear view of how our brain does its job. Evolution is the key.

Any thinking about evolution starts with a question:

What is the such-and-such structure for? If you are considering the evolution of the kidney, you want to know what the kidney is for—that is, to secrete urine (remove wastes) from the body—before you try to figure out how it evolved. The same is true for the brain. Most neuroscientists are, remarkably, unable to answer this question, but evolutionary biologists know the answer right away: The brain is for making decisions about how to enhance reproductive success. That is what the brain is for, no more no less. In its capacity to carry out that task, it can do a lot of other things, which come along for free and are what researchers have come to study while leaving unexamined the reason the brain exists. Yet once we realize that the brain can be explained only in terms of how it handles information and makes decisions, we gain precious insight into mind-brain relations. The brain is not primarily an experience-storing device that constantly changes its structure to accommodate new experience. From the evolutionary perspective it is a dynamic computing device that is largely rule driven; it stores information by manipulating the value of simple arithmetic variables.

Voices of reason are popping up within the psychological sciences to articulate this view. Former empiricists are turning their heads toward evolutionary theory. One of the leading proponents for the necessity of this view is psychologist Randy Gallistel at UCLA. Gallistel has taken the time to master the biological view and has imported his understanding into the heart of psychology. Although trained under the intellectual umbrella of association theories of traditional psychology, he has constructed from the essence of evolutionary theory a solid perspective for viewing the biological substrate underlying normal development and learning.

In marked contrast to notions that the brain is a general-purpose device which can be trained to do almost anything is the deep-seated biological concept of adaptive specialization. Your body contains a constellation of mechanisms organized in a hierarchical way—all built specifically to serve your needs. Rhodopsin, a chemical in the eye, changes light energy into neural energy and information. That is what it does. In contrast, hemoglobin is a chemical that carries oxygen to every corner of your body. The rhodopsin molecule cannot carry out the functions of the hemoglobin molecule, and vice versa; each is adapted to do its own thing.

The language area of your brain can't recognize faces, and the face area can't do language. One recent brain imaging study followed patients who had suffered aphasia following a stroke. After two years some of them were able to use language, but others could not. This kind of clinical finding has usually been interpreted to suggest that another part of the brain took over the language function. The brain imaging results showed, however, that the patients who did not recover showed activation in the right brain. The patients who did recover showed activation around the damaged zone in the left brain. Perhaps the right brain was trying, but it was shooting blanks. The recovery was due to spared cortex surrounding the original lesion.

The fundamental truth about the specialization of molecules has usually been checked at the door when scientists ponder how the brain enables mental function. It is as if the brain isn't part of the biological system! The prevalent view was, and still is in some quarters, that nothing in the brain is specific; it comes from the factory with no built-in adaptations. As Terry Sejnowski claims, it has to be taught everything. And that's why the environment, or nurture, is so important. And, yes, that's why we should read to our babies.

. . .

Why have many brain scientists rejected or at least resisted the fact of adaptive specialization? Well, imagine that you have decided to study how neural tissue learns. This means you are working in a laboratory setting and are confined to the rules of science, which require a beginning, a middle, and an end to an experiment. Sitting in front of you is the nervous system of a "simple" animal. You place an electrode at one point and stimulate the neurons at the tip of the electrode. That is the beginning. At another point you have an electrode that records information from neurons that are somehow connected to the ones you have stimulated. That is the middle. You do this several times and begin to see relations in the responses of recording electrodes. Indeed, the responding neurons appear to be learning something. Voilà, you are at the end. Clearly, nervous systems can learn and be conditioned to do anything!

Think of all those experiments from Ivan Pavlov to B. F. Skinner and on to more contemporary behaviorism as practiced by computer scientists. A once-neutral stimulus called the *conditioned stimulus* is paired with a shock or a food reward called the *unconditioned stimulus*. This ghastly terminology depicts how learning is played out. The conditioned stimulus, say a light, becomes salient to an animal because it becomes associated with the unconditioned stimulus, which is either a negative or a positive motivational event. If it is positive, the animal approaches the stimulus; if the stimulus is negative, the animal avoids it. In each instance learning takes place. This paradigm has been played numerous times, every conceivable variable has been probed, and the laws of association have been established.

There is no arguing that animals can "learn" in these situations, that neurons can respond in the laboratory in an orderly and prescribed way. But it is wrong to claim that these demonstrations of what an animal and its neurons can do in experimental conditions prove that the human brain is idling in neutral until it experiences the world. This massive body of research tells us only how the adapted brain can respond to artificial contingencies; it reveals zilch about the organization of the biological system. Teaching a laboratory cat even the simplest association usually takes hundreds of trials of conditioning. When a cat is moving about in the natural environment, only one trial will teach it that the backyard fence is electrified and gives a shock if touched. Laboratory phenomena indicate little about the fact that the brain, like the body, is a collection of adapted learning systems nested together to allow for specific responses to specific challenges. Laboratory phenomena mask the truths of how the brain is built to acquire information from the environment.

Nowhere is this more obvious than in the study of human memory. Plagued by most psychologists' refusals to start their research programs with the basic question—

What is X for?—research in human memory continually glides off to phenomena that may have little to do with how the brain handles memory. The evolutionary perspective on memory is that its main function is to localize things in space. Where did I hide the food? Where is my cache? Where is the base camp I just wandered away from to find food? Where did I see that cute guy? Think how easy it is to remember where you park your car every day. You can, with no rehearsal, walk straight to it at day's end.

. . .

This kind of memory contrasts sharply with the learning of nonsense syllables, the kind of task encountered in the laboratory. If you were given a nonsense syllable to remember early in the morning, could you easily recall it at night without practicing it? The brain can learn such things, but it does not normally do so. Experimental psychologists might next ask how they can interfere with the learning of these nonsense syllables. They develop a research program and define a concept. One example is the concept of "proactive interference": If a subject performs a second task before the experimenter measures the first task, the efficiency of processing that second task can be affected. Hundreds of experiments showed that interference is there. But when Douglas Medin at Northwestern University applied interference to more evolutionarily grounded spatial memory tasks, he saw no effects. A generation of psychologists over the past twenty years had studied a phenomenon that probably has no relation to how and why the brain normally remembers anything.

What, then, are examples of adaptations, and how do they bear out the evolutionary perspective on brain organization? Gallistel supplies many examples of adaptations gleaned from the zoological literature. My favorite is the behavior of the seemingly jet powered foraging ant (*Cataglyphis bicolor*) that lives on the hot Tunisian desert floor. This beast has the remarkable ability to return to its one-millimeter nest hole from a distance of up to one hundred fifty feet, the length it might travel to find a morsel of food. When the ant starts out on a hunt for food, it crisscrosses along until it finds some. Then the ant picks the food up and returns to its base in a direct shot. How on earth does it do that?

Two biologists, Rudiger Wehner and M. V. Srinivasan at the Australian National University, figured out how by finding an ant loaded up and heading home. They picked up the ant, transported it about half a kilometer away, and set it down, pointed in the correct direction. In unknown territory the ant immediately headed in the right direction. Just at the distance it would have found its home nest, had it not been moved by the scientists, the ant stopped and began to look about for its nest hole.

The ant used dead reckoning to get home, much like mariners use to navigate. The navigator on a vessel records the direction of movement and the speed of travel (velocity) at regular intervals. He then multiplies the velocity by the interval since the last reckoning of direction and sums these intervals to get the net change in position. Substitute an ant for the ship and you see the same method at work.

This analysis is significant for knowing how we should

view what the brain does when it is revealing a competence. The neural net, the assembly of neurons involved in subserving a behavior in the ant (or human) brain, somehow keeps track of the ever-changing value of a variable, say velocity. The brain must hold one value in some kind of memory and then add or subtract from it as the ant progresses to its next position. As these processes go forward, the decision-making system of the ant's nervous system reads the values. When the ant turns to go home, it reads the data from the latest value-holding networks and proceeds on target, like a robot.

The lowly ant strips away all the fancy talk about learning and our notions that experiences are encoded in our brain like a new pair of shoes is placed in our closet—in a box, with a name on it. The ant, via Randy Gallistel, seems to be telling us that the way of looking at the nervous system as a thing seeking to be physically altered by some synaptic structural change is asking more of the brain than it need do. You don't have to beat it over the head with hundreds of conditioning trials, each etching the memory of an experience more deeply into the brain. No, an experience is represented as a set of values in a tightly built structure of neurons. While the nervous system can modify what it puts out, it does not follow that the brain has to alter its structure to carry out the behavioral change.

The animal kingdom is replete with dozens if not hundreds of adaptations that work in one learning context but not in another. As with molecules, each is built to do only its own thing. The vast human cerebral cortex is chock full of specialized systems ready, willing, and able to be used for specific tasks. Moreover, the brain is built under tight genetic control.

. . .

But does the brain *develop* under genetic control? Or is the structure easily malleable—if not during all of life, at least during early development? No one becomes exercised at a report that the fly, ant, mosquito, or grasshopper has a nervous system that develops under firm genetic control. Those pesky little robots of nature must be largely genetically determined. Neuroscientists believe this and are working out the genetic and molecular mechanisms by which these creatures have their nervous system so exquisitely hooked up.

Enter the vertebrates. The rat has a good-sized brain with lots of neurons that process masses of information. For years it has been the animal of choice for laboratory research. How is its brain built?

In the heyday of John Watson, the American behaviorist who held that any child could become anything, many brain scientists, led by the great Paul Weiss at the University of Chicago, believed deeply in the idea that function preceded form. Although Weiss didn't talk too much about central brain development, he maintained that nerves grow out to things like arms and legs and in a largely unspecified form. Then the animal begins to use the arm like an arm, and that trains the neuron to be an arm neuron. So the environment teaches the neurons what they should be and do. The Watson and Weiss views dominated American intellectual life.

My mentor, the ingenious Roger Sperry, who earned his Ph.D. under Weiss, didn't believe it. In several elegant experiments he proved that the nervous system comes knowing what it is doing. He transplanted the peripheral nerves of adult rats by cutting the motor nerve that innervates the left paw and sticking it into the muscle that innervates the right paw. When he stimulated the left paw, it signaled the brain because the sensory nerve from that paw was still intact. But the brain withdrew the newly surgically innervated right paw, not the left paw, which meant the brain had sent the message to the left paw nerve which was now attached to the right paw. There was no training or adapting to this condition. There was no plasticity. The rat did not modify its behavior after it had been surgically rewired—even though the old behavior was creating problems. It didn't change because it couldn't change.

Sperry then tried the monkey, and things became more interesting. He rearranged the nerves in the monkey's arm so when the animal went to flex the arm, it would now extend, and vice versa. After much time the animal seemed to adapt, but it became apparent that there was no real adaptation. The monkey learned to think flexion when it wanted retraction, and vice versa. Its brain underwent no plastic changes.

When his critics asked about the developing brain, Sperry started working on an animal he could study during development: the goldfish. Unlike the rat, the goldfish will regenerate its optic nerve if it has been crushed. Sperry demonstrated that neurons from the retina grow to specific points in the fish brain. When he tried to divert them to incorrect areas, they would grow past these zones and hook up to the correct ones. He couldn't make the neurons grow where they didn't belong. He concluded that brain development is controlled and guided by the precise machinery of genes.

Molecular biologists like Corey Goodman at the University of California, Berkeley, have cleverly demonstrated just how exact the growth of neurons is. He studies simple, genetically alterable animals like grasshoppers. These scientists have confirmed that axons plow through foreign tissue to reach their target areas. They do this by some kind of "molecular sensing" that goes on between the tip of the growing axon and the surface of tissues along the way. There is even evidence that chemicals being released at the target help pull the axons to the right destination. All of this is under tight genetic control. If certain genes are altered, the axons grow aimlessly. In sum, the insect nervous system seems to hook up before there is any meaningful input that might have a modifying influence on it. Goodman thinks some fine connections of the tiny dendrites on each neuron may not be under genetic control. What these minute connections do in controlling behavior is still a mystery.

Sperry believed that all the connections between the neurons and the target are also specified by tightly controlled genetic mechanisms in animals with large and more complex nervous systems. But some neuroscientists think environmental factors play their hand at this point in development. What actually happens here is still not known. We do not even know if the fine connections of the dendritic tree alter the actions of a large neural circuit. But there is strong evidence that disruptions of the normal developmental milieu of an organism can upset the usual overall wiring pattern.

In the early sixties David Hubel and Torsten Weisel showed that the brain comes with its hardware already established. They described with exquisite accuracy the visual system of cats and went on to show that kittens have the same organization, even before they have very much visual experience. These elegant experiments were another piece of evidence for the view that genes control development. Hubel and Weisel's work, which received the Nobel Prize along with Sperry, gave birth to another powerful idea: activity-dependent development.

Hubel and Weisel discovered that the normal organization of the visual system can be changed by simply patching one eye at birth. This alters the neuronal wiring pattern in the visual cortex—instead of most neurons responding to information from both eyes, they respond to information from one eye. Changing the natural activity of the neurons leading from the retina to the cortex modifies the neuronal wiring of the cortex. Normal development is obviously influenced by neuronal activity; hence it is activity dependent.

Carla Shatz and her colleagues at the University of California, Berkeley, have carried out experiments that searched for a key role for physiological activity during the visual system's development. She determined that the broad structure of the visual system is established by genetically driven systems. The fine details of organizaton depend on the emerging activity of the neurons being hooked up. However, it is a strange kind of experience that is called upon. Shatz studies the development of the retina and its connections to the brain. All this goes on before the rods and cones, the key cellular elements in the eye that transduce light energy into neural signals, are even working. Yet the optic nerve, the retinal communication link to the brain, is already being established in the brain.

The idea of activity dependence in development has uncovered a remarkable fact about the brain. Yet it remains unclear how it contributes to the development of the whole brain. For example, information about color, form, and motion is sent up to the cortex from the retina. The cortical organization of this information is stereotyped and precise; it is localized and bundled together in designated brain regions. How does this happen?

Larry Katz and his colleagues at Duke University cleverly devised a way to stimulate retinal neurons by implanting electrodes in the optic nerve. They stimulated neurons so they synchronize nerve impulses in a way not encountered in normal development. Katz wondered whether sending an incorrect electrical message would change the way the visual cortex develops. The answer to his question leads to more questions. Katz found that while the artificial messages mar how individual cells respond, the development of the young brain's visual maps and bundled regions does not change. The known maps that exist in the brain's sensory regions remain the same. They stay where they should be and in proximity to other maps.

The broad scaffolding of the brain is built by genetic mechanisms, which also control almost independently the specifications of what connects to what, but the details of cortical arrangements might be left to experiential effects. However, the so-called experiential effects are merely brain activity, not necessarily encoded information from the environment. Brain activity is not the same as experiential or environmental effects. As Wolf Singer of the Max Planck Institute in Frankfurt has suggested, it is more like the brain uses its own information-processing ability to help build itself—one part generates a test pattern and another synchronizes with it.

This is not to say that experience has no effect on the nervous system. After all, I speak English, not Japanese. I know something about neuroscience, not ancient history. There surely is a kind of brain plasticity that allows us to learn and store information. According to E. G. Jones, Ramón y Cajal, the greatest anatomist who ever lived, remarked at the end of his Croonian lecture given to the Royal Society in 1894:

> The organ of thought is, within certain limits, malleable and capable of perfection, above all during its period of development, by well-directed mental gymnastics. . . . [T]he cerebral cortex is similar to a garden filled with trees, the pyramidal cells, which, thanks to an intelligent culture, can multiply their branches, sending their roots deeper and producing more and more varied and exquisite flowers and fruits.

When it comes to the human brain, Cajal's sentiment and those of others such as Sir Charles Sherrington have been touted as support for the view that the human brain is different—more plastic and less doomed to be what the genome instructs it to be. Some believe, as we know from Sejnowski, that the very establishment of neurons is experience dependent; others claim that only the subtle interactions of the tiny dendritic spines of neurons are open to change and rearrangement. Read, read, read to those babies. The sound patterns will be translated into an ideal neural pattern in the auditory nerve that will enhance cortical zones concerned with reading. Or so the logic goes.

As we explore how the human brain develops, it is important to keep in mind the fundamental flaw in the reasoning of those hoping for evidence of plasticity from experiments that disturb normal development. Evolution, after all, takes place in an environment. Species become what they are because they adapted to an environmental niche. Change the niche, and the species adapts, or dies.

The same is true for the brain. It develops in the physiochemical milieu of the skull and is accustomed to the young nerves it is spinning out being stimulated in certain ways. The genetic mechanisms that guide the growth and patterning of neurons were devised for this milieu. The genetic component in that milieu is huge, so there is no reason not to think it guides most of the details of development. When the milieu is changed by clever neuroscientists who intervene and stimulate neurons in abnormal and bizarre ways, the brain responds to this new niche by doing different things. But it doesn't do them in any directed way. The brain simply responds differently, and hence the resulting networks are different. This response hardly suggests that the brain is plastic in the sense that it has rewired itself.

Yet not to assess a definite view that has become prevalent in some circles of brain science would be wrong. The work builds on some real laboratory phenomena that have captured the imagination of many. Perhaps the leading voice for the view of brain plasticity is Michael Merzenich, the talented and very clever brain researcher at the University of California, San Francisco. Merzenich and his colleagues demonstrated cortical plasticity in a variety of experiments. One of the most straightforward and compelling demonstrations was the description of cortical reorganization of the 3b area of the somatosensory cortex. This area receives input from the hand. Following single or double digit amputation in owl monkeys, it changes its shape. A few weeks to months after the amputation, the topographic representation of the hand had reorganized such that the neurons formerly representing the missing digit represented small skin regions on one of the remaining, adjacent digits. Similar findings have been reported in the primary auditory cortex following cochlear lesions.

This work led other researchers to reexamine amputees' responses to their injuries, a kind of plasticity much discussed in the neurological literature. These patients often perceive a "phantom limb"—they think they actually have their missing limb. Although these amputees were first dismissed as hysterical, careful observations of war veterans indicated that a phantom limb is virtually always perceived initially and may persist for months, years, or decades. If cortical reorganizations happen in humans, then a phantom limb may activate neurons that formerly represented the missing limb by stimulating whatever new skin surface has come to be represented in the deafferented limb area.

V. S. Ramachandran at the University of California, San Diego, reasoned that if this were so, it should be demonstrable. In the normal brain the somatosensory cortex that receives information about the body surface is organized so that hand and face representations are adjacent. If the patients had undergone cortical reorganization, stimulating their face and stump skin might activate regions of the cortex formerly representing the missing limb. This could give rise to bizarre sensations—especially sensations on a phantom limb. This is exactly what happened. Ramachandran studied a young man about four weeks after his arm had been amputated just above the elbow. About four weeks afterward, when a Q-tip was brushed lightly against the patient's face, he felt his missing hand being touched. Indeed, with careful examination a map of his missing hand was felt on his face! In a *coup de grâce*, Ramachandran goes on to report two other cases:

> A neuroscience graduate student wrote to us that soon after her left lower leg was amputated she found that sensation in her phantom foot was enhanced in certain situations—especially during sexual intercourse and defecation. Similarly an engineer in Florida reported a heightening of sensation in his phantom (left) lower limb during orgasm and that his erotic orgasmic experience "actually spread all the way down into the foot instead of remaining confined to the genitals—so that the orgasm was much bigger than it used to be."

Taken together, animal studies and observations of human amputee patients converge on the notion that the cerebral cortex can be modified in the adult, and in a predictable way according to the nature of the manipulation or deafferentation. Because we all know we learn things and brain researchers believe most learning goes on in the cortex, it is tempting to assimilate these ideas and propose that modifications of the cerebral cortex like those described previously are the foundation, indeed, the underlying neuronal mechanism, of skill acquisition. As seductive as this notion is, however, the results don't eliminate the possibility that cortical reorganizations are simply a response to traumatic injury, rather than an extreme example of the normal operations of the cerebral cortex. If the latter is so, it is still plausible that the brain undergoes no demonstrable structural change during normal skill acquisition.

However, the evidence suggests that cortical modifiability is directly related to behavioral modifiability. Another way of looking at findings like Merzenich's is that the cortical reorganizations reflect normal, dynamic processes of the cerebral cortex that allow for the processing of new information related to gaining better perceptual, motor, and cognitive skills and for adaptation to a changing environment. But if we return to the evolutionary perspective and the much-needed knowledge of natural selection to understand how we got here, what would be the adaptive value of having stimulation of the face give rise to a phantom hand? The answer has to be none. Could it be a random event signifying nothing about normal brain mechanisms?

It could be, but the very clever young neuroscientist Greg Recanzone at the University of California, Davis, argues differently. Evolutionary pressures on all adaptations are related to an organism's ability to procreate successful offspring. Certainly, the animal has no prior knowledge of how an adaptation will affect its reproductive success. Thus, there cannot be significant evolutionary pressure for individuals of a species to reorganize the cortex after limb amputation. In real life, an animal that has a leg cut off usually dies within minutes to days. Recanzone notes that if evolution were effective for this sort of problem, we would see a lot more three-, two-, and one-legged animals on our nature walks. Instead of expanded representations trying to make the face or the stump skin do the job the hand used to, such cortical reorganizations are probably an epiphenomenon. The deafferented cortex has nothing to do, and there is no penalty for its doing the job of its neighbors. After all, in a system designed to alter synaptic strengths, which is surely what the brain does second by second, neurons that have lost their normal dominant input will begin to respond to previously weaker, nondominant input that was already in place before the injury occurred. Indeed, in recent studies G. W. Huntley at Mt. Sinai Medical School in New York City saw the sort of plastic changes Merzenich reported only when there was a preexisting underlying neural structure.

Recanzone argues that a system that continuously modifies synaptic strengths can readily change these networks of neurons and maps to adapt to new information and environmental demands. This may actually work if there is a preexisting structure. For example, if seasonal alterations modify the food source, an individual may have to improve its skill—like hearing a beetle crawling under the bark of a tree, recognizing a camouflaged prey, or using a dry stick instead of a green one to catch termites. Once the season changes again and a different food source is presented, the individual can use the skill that worked before. Similar environmental demands, such as avoiding seasonal predators or acquiring mates, would presumably benefit from this adaptability.

Such a system would also be beneficial over a much longer time. Throughout life age causes changes that the sensory and motor systems must adapt to. Older individuals are commonly not as spry as they were when younger, but their greater experience and wisdom should help them avoid dangerous situations by relying on memory and practiced behavior patterns that have proven successful in the past. Thus Recanzone argues that while it may not be particularly adaptive to represent the cutaneous regions of the face in the area of the cortex that formerly represented the glaborous surface of the hand, as seen in Ramachandran's patient, the benefits of adapting to a changing environment far outweigh the potential disadvantages of having a phantom limb. I would add that although the so-called plasticity changes seen following an amputation reflect a mechanism by which the brain adapts to modification, all these alterations could be accomplished by a network laid down fairly definitely at birth.

It is also important to keep in mind the distinction between plasticity and continuing maturation. All sorts of processes continue to develop throughout early life. The challenge is to determine whether the adult brain results from these unfinished processes being established by genetic factors or by experience.

Consider the developing human brain. Since it is a massive structure, I'll limit myself to the huge cerebral cortex. The cortical surface, or gray matter, area is one hundred times larger in humans than in monkeys; the thickness of the cortex in humans and monkeys is virtually the same. The expansion of the surface area seems related to our vastly more impressive mental powers.

During the embryonic phase, which lasts until about the

eighth week of gestation, all the nerve cells that will eventually make up the cerebral cortex wait to be dispatched to discrete zones in the cerebral cortex. From about the fifteenth week of gestation the cortex expands rapidly. Literally billions of neurons leave the cortical plate, which sits below the first layer of cortex and begins to form the brain's gray matter. Miquel Marin-Padilla, an intellectual heir of Cajal who has worked alone in his laboratory at Dartmouth Medical School for thirty-five years, was the first to describe the events, using embryonic and fetal tissue to reveal the unfolding of the cerebral cortex.

From approximately twenty-four weeks of gestation to birth the entire cortex develops so fast that all known adult-type cells are present and functional. This intricate mosaic of neuronal cells, consisting of blood vessels, glia cells, and other tissue, is the same as the adult cortical architecture. This perinatal period is also a sensitive time, when the young brain can be seriously damaged. All too familiar to neuropathologists like Marin-Padilla, perinatal insults like asphyxia can cause permanent injury to the young cortex. The developing cortical cells' wayward attempt to reestablish their connections create a deepening pathological condition that results in several childhood disorders, such as epilepsy. Even at this young age the brain does not fix itself. The concept of plasticity seems a cruel joke.

. . .

Neurobiology has augmented this profile of cortical development by discovering a tremendous overproduction of cells during cortical development. Although billions of cells remain in place after neurogenesis, as many as 50 percent die off during development. This exuberant growth and retrenching has captivated many researchers and led to the postulation that experience or cellular activity determines which cells survive.

One of the first neurobiologists to study exuberant growth was Giorgio Innocenti at the University of Lausanne. He observed in the 1970s that during development the cells making up the long fiber tract that interconnects the two brains, the corpus callosum, comprise many more neurons than seen in the adult animal. Cortical axons grew "exuberantly"—too many grew in a certain place in the brain. Perhaps an experiential event determined which axons survived. A veritable mountain of research now proves exuberant growth in the brain, followed by massive cell death or axon loss. The dendrites, each neuron's structure for contacting other neurons, are also subject to overgrowth and retraction.

At this time we don't know why there is a costly overproduction of cells and axons, why some cells survive and some do not, or whether there are implications for early psychological development. James Voyvodic at the University of Pittsburgh has spelled out at least three plausible reasons for this odd overproduction of cells. The explosion of cell division that gives rise to the cortex produces many "lemons," and these bad cells die off because they are unfit to serve. The problem with this interpretation is that this process doesn't occur in other developmental systems. Another idea Voyvodic proposes is that the massive overgrowth produces a large crop of cell types selected according to their phenotype. The idea here is that the overgrowth provides opportunities for new arrangements and cell types that possibly change the brain's function in a way that may be evolutionarily advantageous to the organism. Finally, Voyvodic proposes that cell durability is predicted by the presence or absence of enough survival factors. This suggests that survival is not cell specific; cells that endure simply received the extracellular chemical growth factors necessary for survival.

The intriguing hypothesis that real-world experience sculpts neurons back from their exuberant growth overlooks a major point. Most exuberance and subsequent pruning happens before birth, leaving moot the possibility that this neural development is under psychological guidance. As Yale scientist Pasko Rakic, perhaps the world's leading neuroanatomist, has stated, the infrastructure of the human brain is in place at birth. It is surely put in place by an extremely dynamic system of growth and pruning and calls on activity-dependent processes throughout development and postnatal life. Genetic control mechanisms play out their role in a dynamic physiochemical milieu that they have come to know and count on to do their job. The visual system, for example, knows skull size is going to change, as will the interocular distance. This system wants a mechanism that can adapt to the change so its final wiring will accurately take into account binocular information and guarantee good stereoscopic vision. Yet none of this cleverness means brain development is not genetically driven.

We still have the issue of postnatal maturation. From the moment of birth all children express their built-in capacities and acquire information at a startling rate. The question Bill and Hillary Clinton raised was whether brains and children can be enriched by experiencing a certain kind of environment early in life. As the *New York Times* stated, "neuroscience" tells us that now.

In fact neuroscience tells us no such thing. It is still unearthing mechanisms that relate to higher mental functions, such as language learning. In my laboratory Jeffrey Hutsler, a young neuroanatomist, followed up on an idea by Marin-Padilla and discovered that the human cerebral cortex undergoes a major thickening starting around eighteen months after birth. Gray matter dramatically expands in the language regions of brain layer II. This anatomical observation runs in parallel with a well-known psychological event. Up to the age of about eighteen months children know but a few words, say fifty. Then they start to acquire up to ten new words per day for several years. By the age of six children have vocabularies of around ten thousand to twelve thousand words. Clearly, young children require exposure to language to acquire these words, but cortical maturation both constrains and drives this amazing process.

. . .

The idea that brain growth and maturation are correlated with behavior is now being considered seriously. When the psychologist Jean Piaget made his seminal observations about the stages of development children pass through, he assumed these stages reflect how children interact with the environment. Jerome Kagan of Harvard University reviewed this and other work, such as John Bowlby's discoveries on attachment. Bowlby, too, considered separation anxiety a consequence of first learning to be attached to a parent through experience. Kagan argues that there is now enough information about brain development to say that Piaget and Bowlby had it wrong. Instead, so-called learned responses reflect continuing maturation of the brain. We don't learn to talk, as most think. We start to talk when our brain is good and ready to say something.

Kagan and his colleagues divide first-year development into two milestones. Big things happen at around two months and then again at seven to ten months. Children at two months begin to lose their grasp reflex—the cute way they grab on to your finger when you place one in their hand. They also experience a reduction in self-generated smiling and spontaneous crying and begin to improve their recognition memory. Could this be due to learning? No way.

Major changes take place in the cortex. Many of these are under the aegis of the brain stem, the older part of the brain that controls motor functions. One mechanism for stopping these events from happening spontaneously might be to have cortical control invade the brain stem. This starts at about two months of age. Neurons in the motor cortex that control the brain stem, the pyramidal cells (they are shaped like little pyramids), madly differentiate and develop. The long nerve tracts, the axons that lead away from cells in the cortex, acquire a sheath around them, called *myelin,* which enables nerves to transmit electrical messages quickly and efficiently.

Even spontaneous smiling is probably controlled by the

brain stem. Babies born without a cortex, the ghastly condition of the anencephalic child, can smile only when their brain stem is intact. As for growth in recognition memory, monkeys' times correlate with the growth and differentiation of the hippocampus, the part of the brain involved with memory.

Kagan maintains that the events occurring at ten months of age (improved retrieval memory, self-initiated locomotion, and such emotional responses as signs of fear during novel events, for example, the appearance of strangers or separation from the mother in an unfamiliar location) can be explained by major changes in brain maturation. Brain circuits are hooking up on a preprogrammed plan, and behavioral changes are automatically occurring in response to maturational shifts in the brain. Genetically programmed brain circuits are doing their job.

. . .

No scientist seriously questions whether we are the product of natural selection. We are a finely honed machine that has amazing capacities for learning and inventiveness. Yet these capacities were not picked up at a local bookstore or developed from everyday experience. The abilities to learn and think come with our brains. The knowledge we acquire with these devices results from interactions with our culture. But the devices come with the brain, just as the brakes come with the car.

Our brains are intricately built, and those precious circuits that allow us to learn are laid down by the genome. Even ephemeral qualities in our lives, like our temperament, are cast by our genetic makeup and provide a big constraint on how we cope with the world. These brain-driven processes influence our decisions, our interpretation of life, and our sense of self. We sit back and watch it play out. Our brain circuits are on the job all the time, doing for us things we thought we were doing for ourselves.

So if you think reading to your kids will make their brains hook up better, think again. This whole episode of politically correct pseudoscience babble reminds me of the late Carl Sagan's dire warning: "It's a foreboding I have—maybe ill-placed—of an America in my children's generation, or my grandchildren's generation . . . when, clutching our horoscopes, our critical faculties in steep decline, unable to distinguish between what's true and what feels good, we slide, almost without noticing, into superstition and darkness."

NOW
AM97

3 THE BRAIN KNOWS BEFORE YOU DO

In our description of nature the purpose is not to disclose the real essence of the phenomena but only to track down, so far as it is possible, relations between the manifold aspects of our experience.

NIELS BOHR, *Atomic Theory and the Description of Nature*

y the time we think we know something—it is part of our conscious experience—the brain has already done its work. It is old news to the brain, but fresh to "us." Systems built into the brain do their work automatically and largely outside of our conscious awareness. The brain finishes the work half a second before the information it processes reaches our consciousness. That most of the brain is engaged in activities outside conscious awareness should come as no surprise. This great zone of cerebral activity is where plans are made to speak, write, throw a baseball, or pick up a dish from the table. We are clueless about how all this works and gets effected. We don't plan or articulate these actions. We simply observe the output.

This fact of brain-mind organization is as true for simple perceptual acts as it is for higher-order activities like spatial behavior, mathematics, and even language. The brain begins to cover for this "done deal" aspect of its functioning by creating in us the illusion that the events we are experiencing are happening in real time—not *before* our conscious experience of deciding to do something.

Many processes that guide us are mental activities, yet they are similar to low-level reflexes in being built-in adaptations fabricated by the brain when it encounters a challenge. We all spontaneously believe we are in control of them, that we guide them to fruition. But often we do not; we simply watch things happen to us and for us. Examples of wired-in mechanisms found in animals are useful in thinking about this fact. Take the spatial behavior of the vole, that ugly little beast living in farm fields. Although all roughly the same size, voles vary widely in how they manage mating. The story of their quirky mating exemplifies how an evolutionary push toward polygamy leads to differentiation in the spatial abilities between sexes. Those male voles that are polygamous can find their way home in the dark, whereas the monogamous females have only limited spatial skills. A cascade of processes that are related and that all go on automatically because of the way the vole's brain is built is responsible for this differentiation.

This curious story, worked out by Steve Gaulin and others at the University of Pittsburgh, builds from a well-known phenomenon: Males perform better than females on some spatial tasks. This is true for humans, rats, and almost everything in between. Gaulin maintains that this cognitive fact is borne out by Darwinian pressures in sexual selection dynamics; that is, evolutionary forces differentiate the behavior of males and females of the same species. Sexual selection pressures usually are not factors, but when they are, it is because the male or the female can enhance reproductive success by behaving differently from the other.

With voles it comes down to polygamous versus monogamous behavior. Gaulin concentrated on two types, the meadow vole and the pine vole. He implanted little telemetering devices into them and carefully plotted how far they wandered from home. Sure enough, the monogamous pine vole stayed around the nest; males and females were similar in their wanderings. Meadow voles behaved differently. The polygamous males wandered all over the place, while the sensible females didn't waste energy running hither and yon. They stayed home. The males needed to find more available mates, which meant wandering greater distances in search of females. When the mating season is over, the meadow vole males' and females' spatial wanderings become more alike.

Once Gaulin established a gender difference in mating strategies that required a more finely tuned spatial capacity for the male than the female, he went on to prove that this superior spatial skill is also present in complex maze learning as done in the laboratory. What prima facie looks like a difference in a cognitive skill is actually a skill borrowed by way of selection pressures to enhance reproductive success. We see, then, how a built-in brain system automatically goes about its job. The tape runs outside of conscious awareness. The part of the brain most connected to spatial processes, the hippocampus, is larger in animals that use spatial cues.

What about more human skills? "Sure those voles are robots, but not us," insist the naysayers. "We willfully construct and control our skills after years of education. We don't use them *before* we know about them. We use our skills *after* we *will* them." Before delving into the so-called "Watergate questions" (What does the brain know, and when does it know it?), let me describe a high-level human capacity that appears to be built into our brains every bit as tightly as the voles' spatial ability—our capacity to count.

Rochel Gelman at UCLA has explored the rich question of how babies, children, and adults of all cultures count. Like the topic of language, this hotly debated issue revolves around whether counting is learned or innate. Gelman, and now many others, view this as a false dichotomy and present evidence to illustrate how innate core capacities can be expressed in different ways at different ages and in different cultures. They show how a universal and automatic brain-based counting system expresses itself differently due to cultural and age variables. And they demonstrate that all members of the human race possess universals.

All humans honor the same counting principles, which always include addition and subtraction. Even when people do simple multiplication, addition comes into play. Our capacity to learn multiplication tables by rote should not obfuscate the fact that if we have to multiply unfamiliar numbers, we usually break them into subunits whose product we know and then add them up. This discovery is borne out by patients who have suffered left-hemisphere damage. Several years ago I studied such a case at Cornell University Medical College. The clinical syndrome of acalculia is complex, but the young girl had a marked inability to add and subtract. These patients frequently can recite multiplication tables because they can memorize them; hence they don't reflect real counting principles. But asked to add eleven and twenty, they fail miserably.

Stanislaus Dehaene at the Institute de la Santé et de la Recherche Médicale in Paris has explored acalculia and come up with fascinating insights. He and his colleague Lauren Cohen have unearthed what all brain scientists like to see: a *double dissociation*. This refers to the capacity for a brain-damaged patient to fail at one task but successfully perform another of equal complexity. Another patient might fail at the second task but not the first. Dehaene and Cohen reasoned that humans engage in adding and subtracting—skills independent of language and rote memorization. If language and computational abilities are truly independent, they should be able to find patients who can count but not compute and who can compute but not count. They found just that. One of their cases could easily count one, two, three, four, and so on but could not add, figure out what was half of four, or do any other simple calculation. The patient was totally bewildered by the mathematics of the task, although he could say all the relevant words. Another patient was able to combine and subtract values, but had difficulty naming the values.

We are sure to find dramatic examples from one culture that violate a principle gleaned from another culture. Karl Menninger maintained that the learning of natural numbers is culture specific. Menninger and others have claimed that some cultures cannot count because they use a small number of words or hand-body configurations for counting, and these do not equip them for large arrays of items. Gelman and her husband, Randy Gallistel, however, have pointed out that this may well be due to a cultural taboo against counting familiar objects like cattle, horses, and children.

Gelman dug up evidence for this notion when she interviewed an Ethiopian family that had migrated from Africa to Israel. Gelman and her colleagues questioned the father; the daughter, who knew Hebrew, translated. The old man used a simple base ten generative counting rule and made it all the way to the thousands. When the Hebrew word for "million" was used, the old man said there was no such word in Amharic because there were not that many things to count at home. When asked about counting children, the daughter interrupted and exclaimed, "You never count children. It's not done."

In New Guinea, members of the Oksapmin tribe count by starting with the thumb of the right hand and going to the right index finger. After the fingers are used, they traverse up the arm, across the shoulders, and down to the left hand, ending on the thumb with the number twenty-nine. This tagging is symbolic and can be used in computations, as shown by Geoffrey B. Saxe of the University of California at Los Angeles. He ran a clever experiment in which he told the villagers that across the island they count from the left hand rather than the right. If a man from across the island has a "left shoulder's" amount of sweet potatoes the way he counts, would he have the same amount as an Oksapmin? All the men quickly answered no. It may be tagging, but there is still computation going on.

Across all cultures the brain has a way to compute, and

it likes to count by adding and subtracting. Stray from those demands and our brains can still do the task, but they do them by inventive means based on simple counting.

. . .

Is there a delay between when our built-in devices go to work and when we become aware that we are engaged in an activity? Can we peek into the brain and see electrical activity starting up and getting things going before we know it? If these wired-in adaptations are lying in wait for the right challenge to be activated, we ought to see activity in the brain before we sense it is going on. Benjamin Libet of the University of California, San Francisco, has been working on this issue for years, and he thinks he sees precisely this phenomenon. Libet tackles two major issues: how long it takes us to become consciously aware of a sensory event and how long neural tissues are active before we will an action to be done.

Curious scientists are allowed in the operating room for certain brain surgeries. During surgery on epileptics, it is frequently necessary to stimulate regions of the brain while the patient is awake but under a local anesthetic. The surgeon can thus figure out which functions are managed by an area of the cortex and determine whether that area can be spared though other regions might be removed. During this procedure the scientists in attendance can learn about the timing of physiological processes and how they relate to conscious experience.

Libet took the opportunity in such a situation to pursue his first question. Starting in the mid-1950s, although not publishing his results until the mid-1960s, Libet tweaked the scientific world with the question, "How long must a stimulus occur before we become conscious of it?" This is the sort of question that intrigues inventive scientists and drives methodologically oriented ones up the wall. How, the latter ask, can you have a measure of when something becomes conscious? Among other things, it takes time to indicate the presence of the "raw feel" after one has it. So how do you get a measure of when the raw feel starts?

Libet began by asking how long an electrical impulse has to be applied to a human's cortex before the person becomes aware of it. When examining the critically needed stimulus in a response, physiologists vary its intensity and duration. The greater the intensity, the shorter the duration. Over many trials one can determine the minimal intensity required to evoke the sensation and how long before the stimulus produces the raw feel. Libet determined the time is five hundred milliseconds, which is half a second between when the stimulus starts in the cortex and the subject reports the presence of a sensation.

Yet we think we are almost instantly aware of stimuli and can react more quickly than half a second. When we are at our best, we can react within a tenth of a second or thereabouts. The concept here is tricky. We have to distinguish between when a stimulus can initiate cerebral processes that move the hand and body and when we become aware of the movement—two different time frames. A single pulse delivered to the skin can evoke an electrical impulse in the cortex in twenty milliseconds. But after that single pulse gets to the cortex, it takes up to five hundred milliseconds for us to become aware of the stimulus. Again, those early signals can begin to recruit other neural processes that might organize a motor response to withdraw from the stimulus. But we aren't aware that the event is going on until about five hundred milliseconds have elapsed. Even though Libet found a half-second delay, we like to believe that we become aware of the stimulus earlier.

Libet's research team came up with the idea that, for subjective time, we automatically refer the beginning of an event back closer to the onset of the stimulus. Think of it this way: If I sneak into your head and stimulate the part of your brain that represents your pinkie, you don't feel the stimulation in your head. You spatially refer it to your pinkie, a full three feet away. Spatially referring events goes on all the time. Libet proposed that we refer the onset of an experience back in time, and he proceeded to test the idea—the sine qua non of good science. First he applied a half-second stimulus directly to the cortex, and then, four hundred milliseconds later, he applied a second single-pulse stimulus to the skin. Amazingly, all subjects reported that they felt the skin stimulus first. Apparently there was no subjective delay for the skin stimulus. Libet carefully observed the cortical response to these two stimuli. There was no locally evoked response to the cortical stimulus train, but the peripheral skin stimulus produced a quick response in ten to twenty milliseconds.

When a stimulus is applied directly to the skin, the neural message from there arrives in the cortex with a bang called the *evoked response*. All the neurons involved in transmitting the information from the periphery to the cortex sum up their action, and out comes the evoked response. But when a stimulus is applied directly to the cortex, there is no big bang, no signal heralding the response from the periphery. Stimulating the cortex directly eliminates this phase of the processing chain.

Libet proposed that this early signal from the skin becomes a reference point for a trick the brain uses to make our sensory experience seem like it occurs instantly. The brain refers the stimulus to the skin back to the quickly arriving local potential. Since that local potential occurs before the first stimulus (to the cortex) has gone through its necessary five hundred–millisecond sequence to become known to the conscious system, it seems to the subject that the second, skin, pulse is actually first. Now that is a big idea, so Libet ran a test on it.

He tested the idea by altering the experiment a bit. Instead of delivering one of the stimuli to the cortex first, he delivered it to the subcortical pathways—a neural system that leads to the cortex from the midbrain. Here, Libet was studying patients suffering from dyskinesias, a set of movement disorders. Part of a therapeutic attempt to help this disorder was the insertion of electrodes into their sensory pathways. This stimulus took a half-second to develop a raw feel, but it also had an early component, just like the peripheral skin stimulus. It, too, had an early marker to refer back to in judging which came first. As predicted, the subject now judged the real first stimulus as the first stimulus and the real second stimulus as the second stimulus.

Libet has provided us with an intriguing possible mechanism for explaining why we think we are doing in real time things that we have in fact already done. His second major observation builds on the work of Hans Kornhuber and Luder Deecke of Germany. They made recordings from the scalp and determined that a certain brain wave begins to fire up to eight hundred milliseconds before a self-paced movement is made. Using another method of recording, Libet determined that brain potentials are firing three hundred fifty milliseconds before you have the conscious intention to act. So before you are aware that you're thinking about moving your arm, your brain is at work preparing to make that movement!

Libet's work suggests that the brain doesn't use the timing of its own firing to represent timing in the real world. When you think about it, this is clever and not too surprising. After all, impulses from the eyes, ears, and cheeks arrive in the brain more quickly than impulses from the distant big toe. We wouldn't want a system that keeps us up to date on all information coming in through time, piece by piece. It would be weird to be aware that things happen in our head before our feet. So the brain makes all kinds of computations that determine when we experience things; we don't just listen to our brain firing directly—any more than we think John is upside down just because his head is at the bottom of our retina. The only thing that counts is information about the world in the brain, not whether the brain *resembles* the world.

This research has raised many tantalizing questions that range across issues of volition, free will, the necessary cerebral events associated with conscious events, and so on. In the present context, however, it tells a significant truth. Major events associated with mental processing go on, measurably so, in our brain before we are aware of them. At the same time these done deals do not leave us feeling we are only watching a movie of our life. Because of temporal referral mechanisms, we believe we are engaged in effecting these deals.

Some scientists find the foregoing a little too exotic for their taste. They prefer more traditional measures and laboratory phenomena if claims are to be made about how the brain does its work before we know about it and whether the brain is continually trying to cover up things. For the moment I'll set aside the intrigue about why the brain plays this charade and turn to an experiment that supports the notion that the brain acts before we know about it.

Scientists agree that it takes about fifty milliseconds to transmit information from the retina to the brain's visual areas. Photons hit the retina, where that energy is transduced into electrical signals that make their way through the retina and up to the visual cortex, rendering any event we see at least fifty milliseconds old. An object we observe may well have changed its position, shape, color, place, or anything else between our initial glimpse and our second registration of it.

Romi Nijhawan at Cornell University recognizes this discrepancy and has unearthed a mechanism that once again provides the brain (and us) with a face-saving device: a dramatic example of how the brain, in an attempt to escape the past, predicts the future of our perceptual world. In his experiment Nijhawan moved a green bar, a line a quarter-inch thick and an inch long, through a black background at a constant velocity. Remember, any instance of that movement is really at least fifty milliseconds old. In one quick instant Nijhawan then superimposed a thin red bar about a sixteenth of an inch wide on the green bar. A static view would mix the colors. Red and green make yellow, and Nijhawan's subjects saw a thin yellow bar on a green background.

Nijhawan wanted to demonstrate that we have predictive perception, that what we see is not what is on the retina at a given instant, but is a prediction of what will be there. Some system in the brain takes old facts and makes predictions as if our perceptual system were really a virtual and continuous movie in our mind. In Nijhawan's experiment the subject interpolates the future path of the green bar that is known to be moving at a constant velocity. When the red bar is quickly flashed, the green bar, from the mind's viewpont, has moved on to another point in space and left the thin red bar behind. The yellow bar didn't travel because the brain knew it was only quickly flashed; there was no need to put it into its predictive machinery. The bar became red again because the green bar, which is in the predictive machinery, moved on without it, leaving it red in its stationary position. This is what is called "nifty" so far as experiments go.

Many experiments underline how the brain gets things done before we know about it. Another example derives from my own research. Our brains are wired in such a fashion that if you fixate on a point in space, everything to the right of the point is projected to your left cerebral visual areas and everything to the left is projected to your right visual cortex. These two separate cortices are interconnected through the largest fiber tract in the brain, the corpus callosum.

If you fixate on a point where on the left the word "he" is presented and on the right the word "art" is presented, you see the word "heart." Your left hemisphere, which is dominant for language and speech, and its visual field are exposed to the word "art." Yet the right hemisphere manages to transfer over to the left the word "he" and insert it in front of "art," so you perceive "heart." None of this is obvious; it just gets done.

Electrophysiological recordings made by Ron Mangun, Steven Hillyard, and myself concentrated on the timing issue and the brain's actions well before our awareness of the consequences. Electrical potentials evoked by stimuli can be measured by using *event-related potentials*, which enable one to track over time the activation pattern of cortical neurons and their cross-hemisphere connection through the corpus callosum to the opposite half of the brain. After a stimulus is presented to the left field, the opposite (right) visual cortex quickly activates. About forty milliseconds later the activity begins to spread to the opposite (left) hemisphere, and after another forty or so milliseconds the information arrives. The word's components are reassembled and, voilà, "heart" appears. Brain processes have done their job before we have a glimmering.

. . .

It is easy to see why clever psychologists began to wonder if formal cognitive psychology had missed the boat. Instead of studying conscious processess scientists should understand unconscious ones! Larry Weiskrantz at Oxford University brought his enormous energy to this idea after he stumbled upon the fact that a patient with a lesion to his visual system and who seemed blind in fact was not. To account for this remarkable occurrence, Weiskrantz coined the term *blindsight*: the ability to see even though the visual cortex has been damaged or removed. Immediately after he reported blindsight, philosophers, psychologists, and neuroscientists became fascinated with the phenomenon.

Weiskrantz's patient had suffered a lesion in the right visual cortex but could, nonetheless, perform visual tasks in what was thought to be his blind field. Moreover, these activities went on outside the realm of consciousness. The patient would deny that he could carry out a task and would maintain that he was responding by mere chance. The unconscious now appeared to be scientifically explorable. This benefit enhanced the excitement within the visual sciences, because subcortical and parallel pathways and centers could now be investigated in humans. Primate studies also flourished. Monkeys with occipital lesions were observed to localize objects in space and carry out color and object discriminations.

As reports accumulated, related issues in other types of brain patients appeared, ready for exploration. Damage to the brain's parietal lobe, for example, causes strange symptoms to appear. If a symptom is on the right side of the brain, most patients experience a phenomenon called *neglect*. Thus, when looking straight ahead, they deny seeing anything to the left of where they are looking, even though their primary visual system is perfectly intact.

This behavior makes it clear that the parietal lobe is somehow connected with attention. Something is happening that is distinct from the parts of the brain that simply represent visual information. Though the information is getting into the brain, it is processed outside of conscious experience. Bruce Volpe, Joseph LeDoux, and I demonstrated this in several ways. We asked patients with neglect to judge whether two lateralized visual stimuli, one appearing in each visual field, were the same or different. Patients might see an apple in one part of their visual field and an orange in the other. Conversely, we might present two apples or two oranges, one in each half of the visual field. The patients performed this task accurately; but when they were questioned about the nature of the stimuli after a trial, they could easily name the stimulus in their right visual field but denied having seen the stimulus presented in their neglected left field.

These were the first investigations in a long series carried out in several laboratories. Taken together, they show that the information presented in the neglected field can be used to make decisions, even though it can't be consciously described. Patients correctly decide that two objects are different, but they can name only one of them.

In the early 1980s my research team began to study blindsight in my lab in New York City. We were fortunate to have a fancy piece of equipment that carefully assesses the eye's position in relation to where a stimulus appears; we could precisely present stimuli within the scotoma, the part of the visual field rendered blind by the cerebral lesion. This expensive machine and the computer attached to it were given to me by Leon Festinger when he decided to leave the field of visual perception and study archaeology. More valuable was his gift of Jeffrey Holtzman, not only because of his scientific skills, but also because he was the funniest man on earth.

We first analyzed a thirty-four-year-old woman who had undergone surgery to clip an aneurysm in her right half-brain. These nasty rats' nests of vessels can break and cause severe brain damage; so when they are detected, they are usually operated on. The surgery was expected to produce blindness in part of the patient's vision because damage would occur to her right occipital lobe, the brain area with the aneurysm. Sure enough, after surgery there appeared a *dense left homonymous hemianopia*; she couldn't see to the left of a point she was looking at. Her magnetic resonance imaging (MRI), a method of taking a picture that reveals the brain's anatomical structure, revealed an occipital lesion that spared both extrastriate regions and the main midbrain candidate for residual vision, the superior colliculus. These intact areas should have been able to support many of the blindsight phenomena commonly reported.

Holtzman started out by giving the patient a simple task. He presented a matrix of four crosses in each visual field and asked the patient to fixate on a point in the middle of a visual monitor and move her eyes to the point in the matrix that flashed. The four crosses were randomly highlighted. The patient had no problem doing the task when the matrices were flashed into her intact field of vision. What Jeff wanted to see was good performance in her blind field. He wanted her to be able to move her eyes accurately but not be able to claim she saw the lights.

Well, the patient was blind as a bat, even though she had brain structures that should support blindsight. Holtzman examined her for months and got nothing. He wrote up the work and published it in one of the finest scientific journals. It met with deafening silence. Blindsight was too big an idea to be shot down by one experiment, even a beautifully executed one.

We left the problem alone for a few years until a new graduate student named Mark Wessinger came to the laboratory and piqued our interest again. Also, my colleague Bob Fendrich came to the lab. By this time we had moved to Dartmouth Medical School and were working with a different kind of patient. Our first case was a New Hampshire woodsman who had suffered a stroke that knocked out his right visual cortex. Nonetheless, he pursued life with vigor and was quite a marksman. Before studying what could and could not be done in our subject's blind visual field, Bob argued that we should do what is called *perimetry*, which employs a complex eye-tracking device, to discover where and how big the blind spot (scotoma) was. For these tests we were armed with a newly acquired image stabilizer, which keeps images steady on the retina despite a patient's eye motions. We carefully explored our woodsman's scotoma by using high-contrast black dots on a white background. In hundreds of trials we presented a matrix of dots in an area of the patient's scotoma while he subjected himself to this trying analysis out of a sense that something might be learned to help others. (It is always heartwarming to see the deep respect that the average citizen pays to the cause of knowledge. Once matters are explained to them, they are invariably enthusiastic participants.)

The efforts paid off. In the sea of blindness we found what we called a *hot spot*, an island of vision. In one small area, about one degree in diameter, the woodsman could detect the presence of visual information. Now if it was truly a one-degree window, a two-degree spot should not be detected. Even though a two-degree spot is larger and under normal conditions would be easier to detect, he should not be able to see it because the black dot would be larger than the window. That is exactly what we found. Follow-up testing revealed that the patient could detect light of different wavelengths in the hot spot. The technology was crucial for the success of the experiment. We could reveal the island of vision only because of the device we had for testing him, and few researchers in the world have access to such equipment. Could the island be the source of so-called blindsight?

Many aspects of our finding directly correspond to the original reports. The woodsman was not confident of his decisions about the lights he could detect. On a scale of one to five, his confidence hovered around the lowly one rating. When we presented a spot in his good field of vision, it was closer to five. At some level, therefore, he was responding above chance but outside of conscious awareness, which is the very definition of blindsight phenomena. But like everything else, the truth is in the details. Our findings indicated that blindsight is not a property of subcortical systems taking over the visual function, because vision was impossible in most of the blind area. Our patient could see only in the islands of vision; the original reports maintained that patients can detect visual information throughout their visual fields.

Another way of examining whether residual vision is supported by the visual cortex or by subcortical structures is to take pictures of a patient's brain and look for spared cortex. We studied our woodsman every which way to find the answer. First, with MRI we could see that part of his visual cortex had been spared. But our information was insufficient because MRI does not tell if the remaining tissue is functional. It might be in place, but damaged or dead.

We followed up with another method that determines if the brain is metabolically active: giving the patient radioactive isotopes. If a cell is alive and active, it will pick up the isotope—an event detected by a positron emission tomography (PET) scan. We tested the woodsman and, sure enough, his spared cortex was alive.

The study of blindsight underlines a general feature of human cognition: Many perceptual and cognitive activities can and do go on outside the realm of conscious awareness. But blindsight need not depend on subcortical or secondary processing systems. The big neuronal blanket called the cortex is full of processes going on outside of our conscious awareness.

Libet's work revealed the unconscious brain. Recall that it takes about five hundred milliseconds for an electrode implanted in the cerebral cortex to elicit the response of the subject being conscious of the stimulation. The same is true for implanting an electrode in the *ventral basal thalamus*, the brain structure that links peripheral input to the cortex. Since this system is part of the touch or the somatosensory system, a five hundred–millisecond pulse creates a tingling sensation. Libet wanted to know if a shorter stimulus, less than two hundred fifty milliseconds, could yield a reliable behavioral response even though subjects maintain they never feel it. He used a simple standard technique called a two-alternative forced choice test. There are two intervals; call them L1 and L2. During one of the intervals he turned on the current. The subject was forced to respond by guessing whether something happened during L1 or L2.

Sure enough, his patients correctly guessed which interval carried the subliminal stimulus even though they adamantly maintained that nothing had happened. The unconscious brain knew. It could make a decision and move the hand to the correct response, and all the while the conscious brain remained clueless. Indeed, most major advances in neuroscience revolve around how the brain is organized to process visual, tactile, or auditory information; the research is carried out on anesthetized animals. The information-processing structures of the brain are alive and well and busy outside of conscious awareness.

. . .

Would you want a new car delivered to you without it being hooked up at the factory? Should this device discover, each time it is given a road test, by divine trial and error how to hook itself up so it works properly? No, you want things to purr and are not surprised to learn that most of the car chugs along outside of your conscious awareness, and its in-built activities go on before you know about them. Yet the doubters, just like the rich and the poor, are always with us. In the next chapter we'll oust that crew by watching the brain at work.

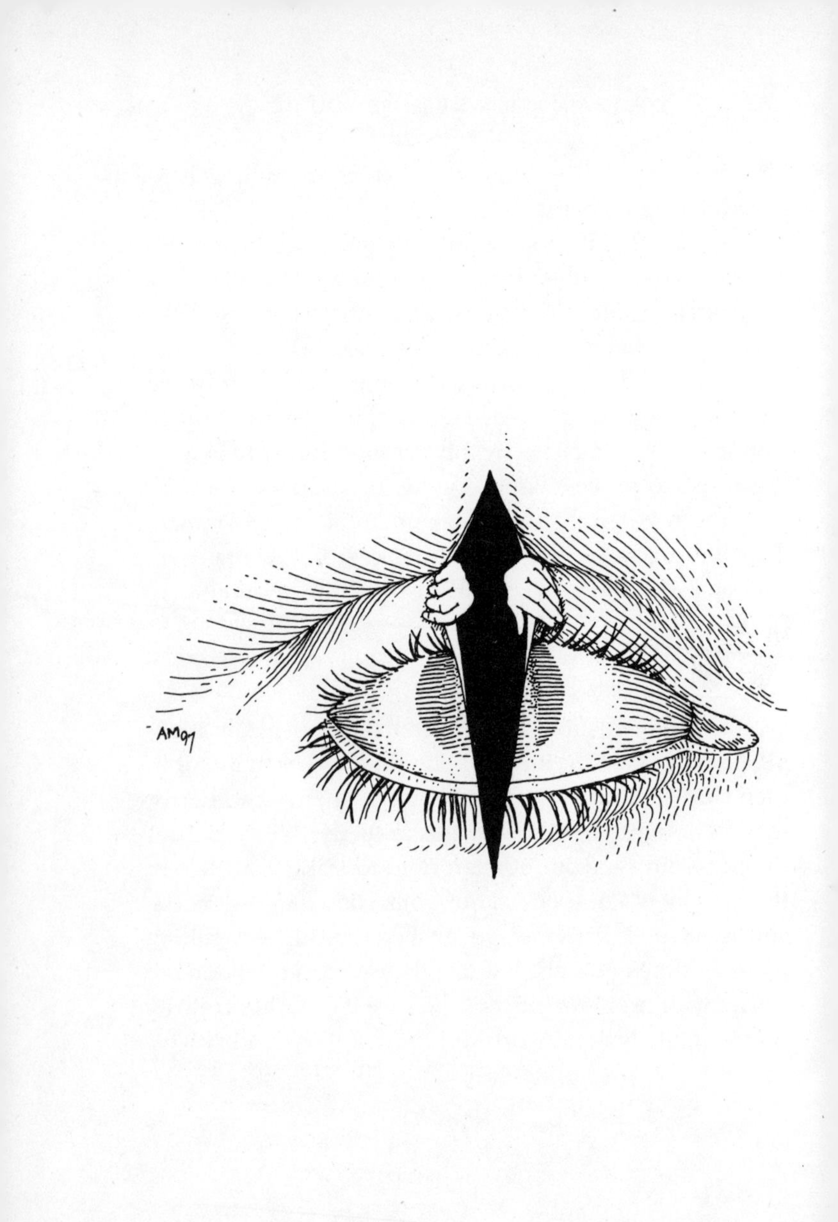
AM97

4 SEEING IS BELIEVING

Nine times out of ten, in the arts as in life, there is actually no truth to be discovered; there is only error to be exposed.

H. L. MENCKEN, *Prejudices, Third Series*

hat you see is not what your retina is taking in. By the time your automatic brain gets done with the electrical signals your retina sends out, you have a highly digested and transformed image. When you "perceive" something, myriad automatic processes have already occurred. One way to prove they are there and doing their job is to trick them, which makes illusions appear. These illusions teach us how the brain works. One frequently sees illusions. By reconstructing automatic events in our perceptual system, I can illustrate how faulty our realities can be.

Our sensory apparatus is only a simple palette for our perceptual and attentional brain systems. Each of these systems is filled with what looks like built-in devices that enable us to perceive the world as best we can and in a way that permits us to maximize our survival. Not only are these devices sprinkled throughout the primary visual system, but they are part of the brain's more ephemeral attentional system. Plenty of perceptual events happen automatically, but there is another level of perceptual awareness. In this later stage of perceptual experience we choose between the array of objects in our view and select some for special consideration. Even this higher level of processing is laced with attentional devices that help us see better.

I have implied that evolution played a large part in vision. Most observers would go along with this view, but this has not always been the case. As with learning theorists, many early vision investigators believed they could link elementary physical quantities to simple mental events. The great German psychophysicist Wilhelm Fechner did not believe vision is adapted to real, spatially structured, typical environments. As Ken Nakayama of Harvard University put it: "Later, the structuralists would attempt to understand perception and higher processes in terms of elementary sensations, as if to build mental molecules out of atomic sensations. Logical and reasonable as it may have sounded at the time, the search for a 'mental chemistry' failed. A description of 'elementary sensations' did not lead to an understanding of perception."

Nakayama, an outstanding visual scientist in his own right, was reviewing the influential life of an American psychologist, the late James J. Gibson. A scientist of unusual talent and brilliance, Gibson stood apart from the structural tradition. He developed a new psychophysics that considered the real environment and its surfaces, light, and distances as the stimuli that brains are adapted to. According to Gibson if you want to figure out how the brain accomplishes its myriad of visual functions, you should be asking what eyes are good for.

Gibson believed an animal's visual system is built to pick up from surfaces information that helps it navigate in the environment. The visual system is not built to represent an exact copy of the actual world; it is built to work by cues that maximize its function. Thus his psychophysics differed from the structural tradition. It was a world of surfaces and textures, not highly quantified measures of the physical dimensions of a stimulus such as a light. Gibson said the brain doesn't want a replica of the external world projected onto it; it simply wants to be cued enough to work right.

Roger Shepard at Stanford University, one of the foremost psychologists in the world, has thought deeply about why we see things the way we do. Inspired by the Gibsonian tradition, Shepard moves easily among mathematics, physics, psychology, and evolutionary theory. He guides us in his thinking about the brain and perception by describing how the brain is built in order to see certain things.

Consider Figure 1. Shepard draws two tables, one in the vertical plane and one in the horizontal plane, so as to add perspective. They appear to differ in shape and size, but in fact they don't. The table tops are exactly the same size and shape. There you are again, watching what your brain is automatically doing for you. If you don't believe it, get out a piece of tracing paper and outline the surface area of the vertical table. Then place it on top of the horizontal table. The two fit exactly. Do it and marvel.

This phenomenon can be explained by how the brain

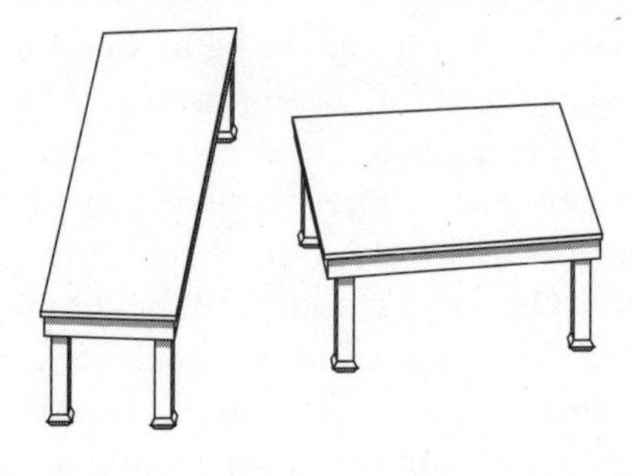

Figure 1

computes information that resides on a two-dimensional structure, the retina, and transforms it into a three-dimensional reality. Here's how it works: Some of the cues that give perspective, the long axis lines, imply that the table on the left is going back in depth. The long axis cues for the table on the right are at right angles to the line of sight. As your brain reacts to these cues, the images of the two tables on the retina are identical. But the brain automatically responds to the depth cues of the left table and infers (for you) that since the table is going back in depth, the image is foreshortened; and because it is foreshortened as a real table in real depth, it must be longer. The same is true for why the horizontal table appears wider. There are also other cues to help create this illusion, but for present purposes, even though you can fully understand that the images are exactly the same and your personal consciousness knows this truth, this knowledge has no effect on your perception. The brain automatically supplies the correction, and you can do nothing about it. In explaining this illusion and dozens of others that vary in interesting ways,

Shepard guides us through the physical universe and its laws of how brains handle information about the natural world. This is not pop or casual psychology. This is the real stuff.

The journey to understanding why the brain automatically processes visual information starts with the nature of the biological system in the three-dimensional world where we have evolved. From Shepard's standpoint our perceptual and cognitive systems were shaped by natural selection in the the same way as our physical size, shape, and coloration were determined. He points out that "a predatory bird has come to have not only sharp talons but also sharp eyes, and a small rodent has come to have not only quick feet but also quick recollection of thc location of its burrow." Each animal has evolved, through natural selection, behaviors and physical characteristics that help it survive the dynamics of where and how it lives. The frightened mouse needs the scurrying behavior to flee the predatory bird. But the two beasts live in a common world with physical laws. Both are adapted to the twenty-four-hour terrestrial circadian rhythm of Earth, to an invariance of physical laws, and—in this case a consequence of a law—to the conservation of angular momentum.

Reaching beyond how natural selection shaped species' capacities to survive local challenges, Shepard wants to know how the invariant laws of physics apply to the mind's many psychological properties. Like the bird's and the rat's understanding of circadian rhythms, creatures of all kinds should comprehend, if sufficiently abstract, general principles about how the brain automatically deals with shapes, colors, objects, and a host of other entities. Psychologists, he thinks, might be looking too closely at how an animal adapts to a local niche. Think big, Shepard implores, and remember that everything adapted to the same invariant laws of the physical universe.

The key concept Shepard is trying to grasp has to do with the role played by invariant laws that guide the workings of what he calls the *representational space* of a domain. Not just humans, but all creatures, large and small, are locked into the representational spaces they possess. Many kinds of representational spaces move up and down the perceptual-cognitive ladder of complexities. One of the core ideas applied to several capacities comes from the well-established laws of generalization, which have been batted around for years in psychophysical and psychological contexts. In plain terms, an organism is confronted with an object that becomes the first member of a class. It could be the letter A or a plant. Then other items are presented to the organism, and it instantly evaluates whether the new item is related to the first member of the class. Values are obtained, and the brain begins to sort things in an invariant way that finds in the set both related and unrelated items.

The extent of the regularity between objects is revealed by *multidimensional scaling*, a mathematical way of placing items so that the distance between points in the space is proportional to the psychological similarity between the corresponding objects in the person's mind. It groups related objects that the brain honors—and does so automatically. Shepard believes that these fundamental mechanisms pertain even to higher-order groupings, such as a species, including humans, who use them, for example, to determine if an object is edible or poisonous. Even more tantalizing, Shepard wants his categories to subsume things like "knife," "bowl," or "chair" for humans or "trail," "burrow," or "nest" for other animals. These mental categories are artifices of the fact that natural selection imbues physical things with such elements as angles, movement, and color.

The tight reasoning and mathematical analysis that go into this proposal are backed up by extensive research. How items fall into what Shepard calls a "consequential space" for an organism—whether it accepts the item as good or rejects it as bad—is too complex for my purposes here. The simple perceptual phenomenon in the turning tables illusion is but one level of a mountain of thought and analysis.

. . .

The human brain contains myriad maps that correspond in an orderly way to the external world. Within the part of our brain governing our visual system there are at least thirty maps. Many of these areas in the brain are designed to handle such special visual tasks as motion, color, and shape detection. This parceling is evident from patients who suffer lesions in these visual areas and afterward display strange abnormalities, such as losing their ability to detect color, shape, and motion. With normal vision the neural processing that goes on in these separate areas is somehow combined so we can enjoy the conscious perception of a colorful moving object like a red cardinal.

When the cardinal sings, we add yet another map from another modality, audition, and map the song to the bird's perch. Again, the brain does this for us automatically, before we know about it. The brain's processor ties the information together into one percept. Anne Treisman of Princeton University has extensively probed this notion that a special processor links the elements of a percept. She discovered the integration of an object's features and a probable mechanism for doing it when she analyzed normal subjects whose attentional systems were overloaded by tasks. When the system is overloaded, it starts to act in odd ways, producing illusions very different from the kind devised by Roger Shepard. For an example, subjects are briefly shown colored letters and asked to report the identity of a simultaneously presented digit, say 4. If they are presented with letters such as a red X and a blue O while doing the digit task, they confidently report that they saw a blue X or a red O!

The subjects' attention was divided and stressed between the colored letters and the digit identification task; hence something called an *illusory conjunction* enters into play. An illusory conjunction results when different elements from different stimuli recombine in the mind's eye to form an object that really doesn't exist. In this example, the red color from the letter X floated over and recombined in the mind to be the color of O. The automatic system, which locates an object in space and adds color, motion, or texture to it, is breaking down. The glue of the perceptual system, the attentional system that usually handles this kind of task and keeps things together in the mind's eye, is stretched to its limits. The brain system that automatically binds features of a stimulus is malfunctioning, even in the normal brain.

A mechanism in the brain that combines the information of perceptual elements must almost by definition be located in areas outside the brain's primary and secondary visual areas. The chef makes the soup, and the ingredients of the soup are discrete entities before they are combined. It is not surprising, then, to discover that the *parietal lobe* is specialized for this task. This is the area that is so powerful in higher-order automatic processes. Stacia Friedman-Hill, Lynn Robertson, and Anne Treisman came upon a most unusual behavior while working with R.M., a patient with bilateral lesions to the parietal lobe. Someone with such damage should have problems with spatial perception; it has been known for years that those with damage there have difficulty with spatial processing. No one appreciated, though, how this part of the brain brings color and shape together in one location.

This patient had lost his automatic binding ability. Indeed, when R.M. looked at objects, he would frequently see color floating astray from them. Presented with two colored displays, he had a high rate of illusory conjunctions—an amazingly robust effect. In normal subjects the stimuli have to be flashed for a very quick two hundred milliseconds to produce effects. R.M. had illusory conjunctions even when the stimuli were left on for ten seconds. The part of the brain committed to triggering higher-order automatic binding was forever damaged.

Treisman and her colleagues have developed many visual tricks to unravel the visual system's organization. By pushing the system to its limits, or by studying patients with relevant damage, researchers can discover higher-order automatic processes. Now that the automatic brain is being aggressively investigated, we are learning more about how it operates. As we ascend into higher cognitive mechanisms, we assume that built-in brain processes are less active. Maybe low-level processes like those that stick color together with shapes, which seem so automatic and ubiquitous, are built in, but activities associated with judging what we can and cannot attend to? No way. These are under our direct conscious control.

In a wide-ranging attack on how we are beholden to other automatic aspects of our attentional system, Patrick Cavanagh, an extraordinarily talented psychologist at Harvard University, and his colleagues started with the question of whether or not our attentional system is limited in what it can resolve in the physical world. We know we're limited in our ability to resolve fine details in a visual scene. Lines printed on a page of a certain width can be only so close together or they become one line.

Well, it turns out that the system which monitors the visual array and tries to enhance how well we process that information also has limits to its resolution. Consider Figure 2. Fixate on the + and don't move your eyes. Notice how easy it is to see the three lines on the right. These lines are well within the capacity or resolution power of your attentional system. Now concentrate on the multiple lines on the left while keeping your eyes fixated on the +. Try to attend to any one of them or count them while fixating. It's impossible. The attentional system cannot differentiate these details; it has its limits, and there isn't much you can do to compensate for them. This higher-level process, which can be applied to a visual scene, has its limits, too. Hence at any moment only part of the visual elements of a scene is available for conscious perception.

This elegant observation opens up numerous issues. The

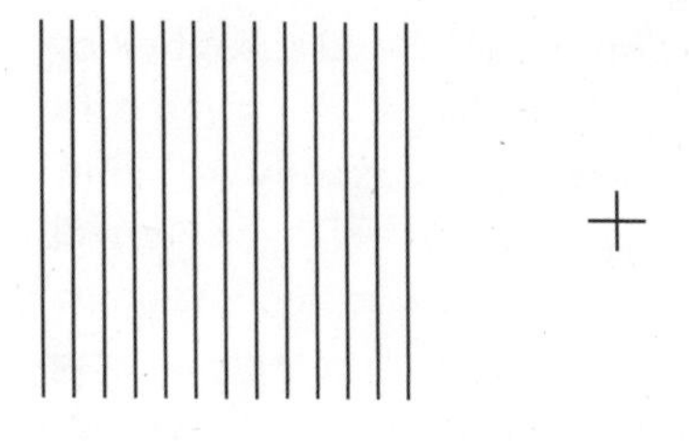

Figure 2

first cortical station for visual information, an area called V1, has the resolution power to see all the lines distinctly. Under ideal conditions the human fovea can see up to 110 black and white stripes painted on a fingernail if held at arm's length. From a sensory point of view, then, the fine detail is in the brain and represented in enough detail to be seen. Yet the attentional system, managed by an area deeper in the brain, cannot pick this information off the sensory surface and use it in a conscious way. Cavanagh wanted to know if the information was not reaching consciousness and was not influencing perception at all or was being used in some subtle way. To explore this, he examined a well-known phenomenon called *crowding*.

Crowding effects are commonly studied by using strings of letters that either make up words or don't—it doesn't matter. When the letters are presented to one side of fixation, those in the middle are harder to detect than those at the beginning or end of the string. Even though subjects can view the letter string for a long time, there is a limit to how much of it they can resolve. Several reasons have been given to explain this curious effect. Some propose that cells in the retina inhibit neighboring cells and disturb the flow of neural information to the brain.

Cavanagh believes this puzzlement is readily explained in relation to *attentional resolution power.* The idea here is that just as the visual system limits our acuity, the attentional system limits how much information we can grasp in any given visual stimulus. He borrows a version of the crowding paradigm and demonstrates how much of this gating of visual information happens outside of conscious awareness. Instead of using letter strings, he draws on a technique called *selective adaptation.* Here grids of lines in one orientation are presented to an observer. After the observer looks at the grids for awhile, neurons in the primary visual cortex adapt to the view. Subsequent presentations of grids oriented in the same way are more difficult to detect, as if the neurons which pick up on, say, vertical lines become fatigued. Yet when we show grids whose lines are oriented in another direction, the neurons' attention perks up, and they chug along at their normal detection speeds. Nifty!

In his key experiment, Cavanagh adapted subjects to either one grid in one orientation or to a set of five grids, which resembled the crowding example. The question he wanted to answer is simple: How much of the information that gets into the brain via the sensory system can actually reach conscious awareness? Cavanagh started by presenting only one grid. Each subject clearly saw the grid and noticed the orientation of the lines in it. After the subject had studied it awhile, another grid was presented in the exact same orientation or in a different orientation. As predicted, the second grid in the same orientation was more difficult to detect because of selective adaptation effects. Those neurons were fatigued. A second grid in a different orientation was readily detected. The subject easily described all the events because they happened well within the window of attentional resolution.

But what about the other condition, when subjects see an array of grids crowded together? Now they are unable to report the orientation of the critical grid. When asked its orientation, the subjects respond at random. However—and here is the crucial part—these same subjects are still selectively adapted to the grid and its orientation. They, too, respond with a greater inability to detect the grid they have already seen, even though they were unconscious of its orientation.

Enter Sir Francis Crick, the co-discoverer of DNA. Though earlier in his scientific career he'd busied himself at Cambridge University with figuring out molecular structure, he decided to spend his later years doing brain research at the Salk Institute. Timidity not being one of his many attributes, he now tries to help neuroscientists pick their way through big issues, like how the brain enables conscious experience. Crick and his neuroscience sidekick, Christof Koch at Caltech, have proposed that this kind of experiment proves our conscious brain cannot be aware of information processing carried out in V1. Crick and Koch have a program aimed at determining which parts of the brain are not involved with conscious experience and which parts are. Their reasoning is clever. Since neurophysiological research on animals has indicated that the detection of line orientation only begins at V1, and since this is the likely site for orientation detections that prompt adaptations, and since in the crowding condition subjects cannot report the grid's orientation, then activation of V1 neurons is not sufficient for conscious visual experience. The message for those in the brain science business is that they should look elsewhere for the part of the brain responsible for conscious awareness. Fair enough. At the same time, it's clear that the unconscious brain, with its millions of automatic processes constantly going on, does use the information—it just happens beyond our conscious control. Crick and Koch's work also demonstrates that unconscious processes are carried out in the cortex, just as we saw with blindsight.

Cavanagh wants to push the unconscious process even deeper into the cortical processing chain. He believes the attentional system, with its automatic capacity to limit what gets consciously experienced, comes into play late during the processing sequences that manage our perceptual world. Many before him have argued that attentional selection works on information within the primary visual system, but Cavanagh thinks not. To arrive at this point he takes advantage of still other crafty visual tricks and combines them with emerging clinical data on patients with damaged parietal lobes, data that neatly bolster his hypothesis. Consider the following.

While it is well known that our visual acuity falls off dramatically as we move away from a point in space, there is no evidence that acuity differs between the upper and lower parts of our visual field. The sensory information the brain receives for visual detail above and below a fixed point in our visual world is identical. But the automatic attentional resolution system treats these two parts of our visual world quite differently. Imagine a rectangular box full of nine constantly moving dots. You are instructed to keep track of two of them as they move about. After a few seconds the array freezes, and you are required to pick out the two dots you were instructed to follow. You can do this with far greater skill if the visual array is in your lower visual field. If the array occurs in your upper visual field, you'll be lousy at the task. What gives? As this and numerous other attentional-based tasks bear out, the lower visual field is more sensitive than the upper field.

The result invites speculation. When did you last spontaneously look upward? We rarely do. That's why, in guerrilla warfare, the safest place to hide is above, in a tree, or, if you are the hunter, to lie in wait above the ground. We never look up. Perhaps we don't because, as we walk along, it is much more important to keep track of the changing terrain. Through evolution we have tuned our attentional system to be more sensitive to objects in our lower visual field. This enhanced capacity to process information in our lower field is consistent with there being more connections to the parietal lobe from the part of the visual brain that represents the lower visual field. Recall that the parietal lobe manages spatial attention. Those connections may have become prevalent over the years through natural selection.

The attentional system has limits to its ability to resolve whether two events at different times are indeed two different events. The temporal sensory system seems good if the visual system can detect a light flickering at fifty hertz. But it cannot tell if the light went on and off when it flickers faster than four to six hertz. The movie and video industries depend on this limitation of our brains. When one stimulus stops and another starts, the events have to be separate enough in time so we don't perceive them as the same event. If they have different spatial locations, then by judiciously separating their timing, the brain interprets the two events as a continuously moving one.

Vision and attention are the processes that are literally and figuratively in our face. The adaptations that assist us in seeing and attending to relevant stimuli in our world are mainly automatic. We can see them run off their processes like a taped message. When it is being played, we observe and collect the products for our conscious minds to interpret. Even though we sometimes have strange interpretations of them, our brains aren't easily fooled—only our minds are.

5 THE SHADOW KNOWS

The most beautiful thing we can experience is the mysterious. It is the source of all true art and science.

ALBERT EINSTEIN, *What I Believe*

ho isn't thrilled by seeing an osprey swoop down and majestically catch a fish in a river? With laser accuracy these birds effortlessly fetch food for their brood. They spot the fish from hundreds of feet up, and before zooming in on it they somehow correct for the water's prismatic distortion between the fish and its appearance from on high. The osprey flies back to its nest, feeds its young, and rests.

All of this goes on automatically. From the eye to the talons, a computation facilitates the kill. This can't be good ol' Kentucky windage at work; development helps the bird to reckon the fish's position. Even though the osprey has not been taught Snell's law of reflected light, it knows that the fish is actually behind where it appears to be. The osprey must estimate not only its own velocity, but also the position and velocity of its prey. And, given water's refractory quality, the bird must take into account the fish's depth. It does better with slow-moving fish than with fast ones, and it hits the bull's-eye better when the water is shallow and clear—all factors that imply limits to how much visual information the bird acquires.

Things become even more devilish as you wade into the details. Not only is there the problem of reflected light from the water's surface; the bird also has to see beyond that to the reflections upwelling from objects in the water below. For seeing through deep blue water it would be good to have a retina sensitive to short wavelength light (425–525 nanometers), and for shallower green water vision at another range would be better (520–570 nanometers). Surface light is reflected at wavelengths greater than 570 nanometers. How can the bird cut the glare of the surface's reflected light so it can see below? Last time I looked, Armani hadn't come up with sunglasses for ospreys.

A prevalent notion is that reducing glare has to do with how many oil droplets are in the bird's eye. Actually, the drops in the retinal cones contain dissolved carotenoid pigments that may be chromatic filters. So the bird has built-in sunglasses. To test this idea, scientists sorted birds according to whether they had a few or many droplets in their retinas. Birds with small amounts might be different; perhaps they hunted where they didn't need to filter out reflected light. Birds with lots of droplets might fish from above the water. This is exactly what was discovered. Razorbills and cormorants, birds that pursue fish underwater, have few droplets. Birds like seagulls and terns are loaded with them. Mother Nature is wonderful, and the scientists who figure things like this out are even better.

Just as automatic instincts are responsible for the osprey building its one-ton nest, which serves as home to it and its children, so they are responsible for the bird navigating to and from its nest during migration season. The osprey is chock-full of neural circuits that automatically facilitate discrete behaviors like nest building, fishing, and migrating. It knows how to automatically correct for the fish's displaced visual image and to place its talons behind where the fish seems to be. The lowly osprey's visual system has an automatic brain unfooled by appearances.

Smart people in corporations put automatic brains to work for them. Union Oil Company had a problem: Its hundreds of miles of gas pipelines across the desert would spring leaks. How were they going to find them all? They injected into their gas lines ethyl mercaptan, a substance that smells like rotting flesh to the turkey vulture, which has a wonderful sense of smell and construes this odor as carrion—a fine dinner. (To humans, curiously enough, ethyl mercaptan smells like urine after you have eaten asparagus.) The vulture relies on its olfactory sense to detect food at long range without seeing it. With its huge olfactory brain and sensitive nasal epithelium the bird smells lots of things. So, you guessed it: Union sent out patrols to find the vultures that sniffed out the leaks.

When thinking about the human brain, it is always worth turning your attention to the ubiquitous automatic processes in animals. Special devices abound in nature, and they perform miraculous feats. We, too, are a collection of devices. Even though our sense of purpose and centrality of will are foremost, there dwells within us an automatic and highly specialized machine.

. . .

The turning tables illusion in Figure 1 proves that the automatic brain creates powerful illusions that prompt the motor system to respond to visual events. Yet if we started to reach for the long illusory table or the fat sideways table, we might be in for a surprise. Things we assumed would be at a place in our visual world would not in fact be there. Our automatic brains aren't fooled by illusions. Consider the Ebbinghaus illusion, named after Hermann Ebbinghaus, a German psychologist who studied visual perception and memory. The illusion on the top in Figure 3 creates the impression that the two center disks are different sizes, but they are exactly the same. The illusion on the bottom has been altered, so now the center disks look identical. To create an appearance of sameness, the one on the right has been drawn larger.

Researchers have known for years that we judge size in a funny way. We depend on perceiving the relative, not the absolute, size of objects in an array. Changing the actual size of an object has little effect on us; illusions abound when we watch TV or a movie. Changes in an object's real size and distance have only a negligible impact on our perceptions.

Melvin Goodale and his colleagues at the University of Western Ontario wondered if the brain is organized in one way to perceive illusions and in another to respond to illu-

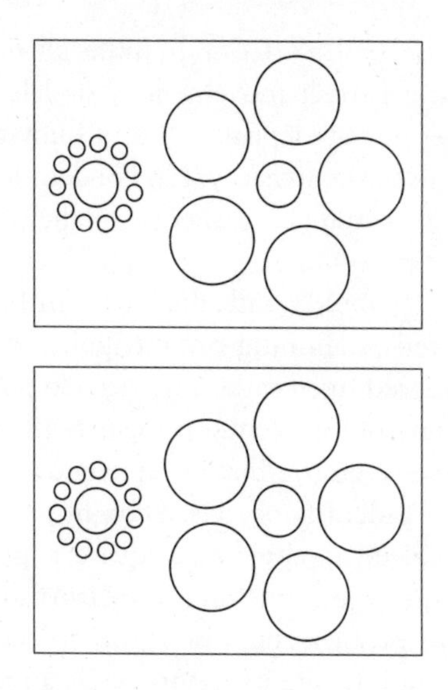

Figure 3

sions. After all, when you reach out to pick up an orange or grab a branch, just as a monkey does as it swings through the jungle, you have to position the opening of your hand according to the object's real size, not the illusory one. Missing out on food or landing on your keister does not bode well for survival.

Goodale has tackled one of the brain's central tasks: transforming a two-dimensional representation of visual information into a three-dimensional motor response. He studies how the brain easily and automatically makes all those computations. The elephant's trunk has about fifty thousand different muscles, each with its own innervation. When the elephant spots a peanut in your hand, how does it create the know-how to orchestrate all those muscles into a swift and perfect movement? Goodale's fascinating observations—not of elephants, but of humans—added a new dimension to the mystery. His concern is how people reach for items that appear larger or smaller than they actually are.

Goodale has grappled with this problem for years, ever since he started examining an intriguing patient, Case D.F., who suffered from carbon monoxide poisoning. It is odd but not terribly uncommon for patients who survive this trauma to develop a bizarre perceptual system. Suddenly, they can't identify objects, a condition referred to as *object agnosia.* Place an apple or a square peg in full view, and, though they see something, they can't identify it. It's not a language problem because if the object is placed in the patient's hand, where its features are transmitted to the brain by touch, the patient immediately names it.

These same patients who cannot see an apple, who are consciously unaware of the object's nature, can pick it up. Moreover, their hand is positioned in the correct manner as it moves toward the object. In more exacting tests, blocks of varying widths are placed in front of patients, who properly anticipate the block's dimensions by adjusting the width between their index finger and thumb as they approach each block. If the patients are not reaching for the object, but positioning their thumb and index finger to estimate the block's width, they can't do it. At the level of conscious experience these patients have no awareness of the object's nature. At the level of visual-motor capacity, however, the brain knows exactly what to do with each object, even though the patients are seemingly unaware of it.

This phenomenon is best understood by stepping back and learning how the brain distributes information throughout the cerebral cortex. As Yogi Berra said, "When you get to the fork in the road, take it." The brain knows to choose a fork, but we don't know how it does it. A valuable clue, detected over the last twenty years, is that once visual information comes into our first and primary cortical zone, the occipital cortex, it goes in one of two directions (see page 117). Something called the *dorsal stream* heads up toward the parietal lobe. The other stream of information, the *ventral stream*, projects into the temporal lobe. Well, the visual system listened to Yogi and took both streams, though inevitable duties to certain information may favor one stream over the other.

Patients with lesions in the dorsal stream, lesions that damage the parietal lobe connections, have little trouble seeing, but a lot of trouble reaching for objects they can see. It is as though they cannot use the spatial information inherent in any visual scene. They know what the object is, but not where it is. That was the classic way to think about things in neuroscience. Patients with lesions in their temporal lobe, lesions that affect the ventral stream, behave like Case D.F. They are agnostic on what an object might be; nonetheless, they can reach for it accurately and with the correct hand posture. Not surprisingly, the ventral stream became known as the "what" pathway.

Against this backdrop Goodale advocates yet another way to think about things. Instead of sorting functions by what and where, he proposes what and how. As David Van Essen at Washington University maintains, the temporal lobes process object-centered information such as shape, color, and size. This part of the brain is all tooled up to identify things. Yet there has to be a place for viewer-centered processing—that is, "What do I do with this thing?" That place is the parietal lobe and hence the dorsal stream. From the viewer's standpoint, the object is constantly changing its position in space. Something has to keep track of it and what it means for future actions.

Goodale's experiment on illusions now seems clear as a bell. Sure, there is a perceptual illusion, one that can be as big as life, as you can tell from Figure 1. But the part of the brain that constructs the illusion is down the ventral fork in the road, miles from the part that generated a movement. This latter part calls upon the actual representation of the image, not the illusory one. In short, the zany misperception doesn't change the dorsal stream's realistic view of the object.

The parietal lobe is the unsung hero of the brain. Scientists like to study visual lobes, if only because things are usually easier to study before they get complex. Frontal and temporal lobes get a lot of attention because lesions here produce behavioral problems ranging from serious language and thought disorders to poor memory. But right inside what amounts to the brain's situation room is the parietal lobe. It is taking the sensory percept and organizing the body's response to it. What could be more fundamental?

When you delve into the parietal lobe's function, the question of timing comes up again. When does the parietal lobe do its work? Indeed, does the automatic brain decide to do things before we decide to do them? It does, and the proof is about as clever as it gets in the neuroscience business.

Michael Platt and Paul Glimcher at New York University look at monkeys' parietal lobes and see how neurons in the parietal cortex respond to challenges. Each neuron has a receptive field, and they record from a group of neurons in the brain area called the *inferior parietal lobule.* Fix your eyes on a spot on the wall next to you. This is just what the monkey does, sitting in a chair. Imagine someone moving a stick all around the wall. There will be a place on the wall, let's say five inches over and five inches up from the spot you're staring at, where the neuron fires at a high rate. Anytime the stick is in that zone, the neuron sounds like a machine gun; and it stops firing when the stick is outside the zone, outside the neuron's receptive field.

When the monkey is looking straight ahead, the experimenters present a task—either look at a light that falls within the visual field of the neuron they are recording from or look at a light that falls outside this field. The fixation point is a light that turns either red or green. If it is green, the monkey is supposed to attend to the light that falls within the neuron's receptive field. If it is red, the monkey should look at the light below it. After that phase of the trial, the monkey waits for the target light to dim; when the light goes out, the monkey is to move to the light at the indicated position.

Platt and Glimcher discovered that a neuron fires most when the stimulus inside the receptive field is the target for eye movement, the place where the eyes should move to. That same neuron does not fire as much if the target is the other light. When the identical target for a potential movement is simply playing the role of distracter, and the distracter conveys the information about when to move, the neuron responds weakly. This means that the neuron is not simply allocating attention to a stimulus, because attention has to be constantly paid to stimuli; the neuron is also encoding information about movement to that place.

Next Platt and Glimcher asked if this neuron knows anything about the probability that the light in its field is going to be the one to move to. They also wanted to know if the neuron knows anything about the size of reward the movement will bring to the animal. In these sorts of experiments reward amount is usually varied by how much fruit juice the subject gets on any trial. The juice-loving monkeys were big on learning how much the parietal neuron knows about the decision to move. Platt and Glimcher varied the probability and the amount of payoff because classical economic theory and psychology's rational choice theory predict what any decision system responds to. Everyone wants to know the chances for a payoff and how much it will be. Platt and Glimcher gained insight to this issue, plus a whole lot more.

They recorded from the neuron throughout the task and arranged their data in several time periods, the first being when the monkey is looking at the fixation light. This is truly a time of uncertainty. The monkey does not yet know to which of the two targets it will move its eyes. Platt and Glimcher added two more time frames: after the target is identified and when the eyes move. They thus knew about the neuron before the monkey decided on which target to fix. They also knew when the target was identified before the eyes moved and the time of the eye movement.

The neuron knows the probability of a light in its field being the target light. The greater the probability of payoff, the more the neuron fires. It fires more during early phases of the trial, when it is uncertain where the animal will have to move. As the trial progresses, the neuron shifts gears and begins to encode more information about the movement to be made. Platt and Glimcher knew this because over many trials they had examined how a neuron fires when it is highly probable that the target is in the neuron's receptive field. They saw how the neuron behaves when the target light is unlikely to be cued. Under both conditions, however, the neuron fires rapidly just as the animal begins to make its move.

Platt and Glimcher went on to evaluate whether the neuron knows anything about the size of the reward if it makes the correct movement. It does. This is some of the first work to reveal that the parietal lobe is truly crucial, participating in what goes into the decision to act. The neurons appear to be tied to the utility of the movement, as opposed to simply attending to a part of their visual world. A lot of this work apparently happens before there is even a hint that the animal is deciding what to do. The automatic brain is at work once again.

Before leaving this topic I must address the nasty question of why use the single-cell approach. There is something deeply unsatisfying about thinking a single neuron determines the choice of behavior. Surely, some kind of complex network does that chore. Why should measuring the activity of one cell tell us any more about the brain than interviewing a New Yorker to learn about the Big Apple?

The only answer is that what can be done now is yielding nifty results. Good science always stretches for more. Michael Shadlen of the University of Washington does this by asking and answering a variant on the still serviceable lightbulb joke: How many neurons does it take? That is, how many neurons need to participate in a decision? After performing a clever mathematical manipulation, Shadlen reckoned about a hundred. I leave the subject with that teaser . . . and the clear feeling that the single-cell approach has much to tell about higher-order brain mechanisms.

. . .

It is not hard to understand why certain visual phenomena have fascinated curious minds for centuries. As far back as 1593, G. B. Porta, a physicist, wondered why we see only one image with our two eyes. The eyes are separated in space, and each has a slightly different angle for looking at anything. This means that the image projected from each eye to the visual regions of the brain should be slightly out of register. Indeed they are, which raises another question: Why don't we see double? How is our automatic brain solving this problem?

Porta explained that we learn to suppress the information from one eye, which clearly happens in the case of the blind spot we all possess in each eye. In cases of strabismus, the clinical state where one eye truly deviates its gaze from the other, people learn to stifle the image from the deviated eye. I have a large blind spot in my left eye that leaves me unable to read easily with that eye. Tests on me show that I, too, suppress the visual input from my left eye.

On the other side of the suppression theory is the fusion theory. We don't usually see double because, among other reasons, we are seeing the same image, but slightly shifted by a quarter-degree or less. This small amount allows for a three-dimensional capacity. According to fusion theory (not to be confused with the hapless cold fusion theory), if the two images are shifted a quarter-degree or less, the visual system fuses them into one. Suppression theorists counter with the question, "How do they know this?" Maybe we alternate the eye being used and the eye being suppressed.

Experimental psychologists immediately began to work out the laws that govern such phenomena and to articulate the underlying mechanism. What happens, they ask, if the images in the two eyes are truly different? What happens if they are unlike in a way that could be additive? They have performed many experiments and drawn some conclusions over a long time. Indeed, back in 1858 P. L. Panum, one of the great psychophysicists, offered rules for when rivalry is seen and when it is not. *Rivalry* is a state in which stimuli presented to the two eyes are different, for a while we perceive them one at a time, and then suddenly, for no apparent reason, we see the other image. The images flip back and forth as we continue to view them. Panum came up with four straightforward principles:

1. If the contours of the stimulus in one eye conflict with the contours of the picture in the other eye, there is great rivalry.
2. If each eye has a stimulus with contours that don't overlap, we see the contours concurrently.
3. If a picture in one eye has many contours and the other eye is seeing a plain black image, we always view the contoured stimulus.
4. If each eye sees a different color, sometimes the colors mix.

This classic psychophysics reflects great care and precision. Yet statements about how our visual system behaves do not tell us how it works. Enter Nikos Logothetis, a brilliant neuroscientist from MIT and now head of his own Max Planck Institute in Tubingen, Germany. Logothetis and his colleagues look into the brain of the multipurpose monkey, whose visual system is similar to ours and who behaves like us on all kinds of perceptual tasks. Logothetis, now at an explosively creative part of his career, beautifully exploits this similarity.

For years scientists assumed that the incongruent images presented to each eye alternate in being "seen" because of fatigue or suppression of the visual mechanisms responsible for processing the images from each eye. (See Figure 4.) Logothetis started his hunt for the brain process that might support this assumption by examining visual system areas V1, V2, and V4, all early processing sites. Recording from individual neurons, he did find a small proportion of cells that seemed to fire only when the monkey thought it saw the stimulus projected to one or the other eye. But the vast majority of the cells appeared to be uninterested in which eye dominates perception. This puzzling response prompted Logothetis to look further into the visual system for answers.

First he performed an extraordinary perceptual experiment that required high-tech innovations in binocular experiments (those on how information from our two eyes becomes coordinated). An alternative belief held that perhaps rivalry is created not by the suppression of one eye channel over the other, but by the competition among stimuli somewhere in the brain. Before searching for the

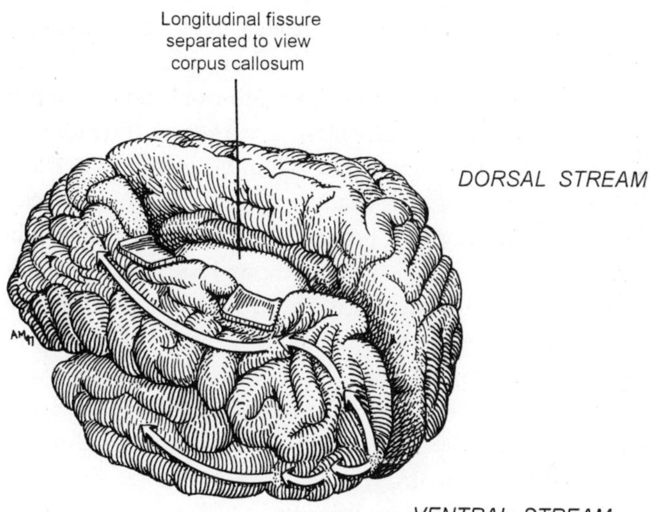

Figure 4

brain's side of the story, Logothetis asked whether we can characterize the rivalry in perceptual terms.

Logothetis and his colleagues devised a way to switch the stimuli presented to each eye without a subject noticing it. Let's say, for example, that the left eye sees a plus and the right eye sees a zero. Normally, we see the plus for a while, and then we spontaneously see the zero, and so on; the switching back and forth is outside our conscious control. Logothetis switched the two stimuli back and forth between the eyes so fast that the subjects were never aware of which eye was seeing what. Thus when a subject was seeing a plus, it was actually being presented half the time to one eye and half to the other. In the mind's eye the sensation was that it had not moved at all and that it was dominating the rivalry. This means that binocular rivalry is due to mechanisms other than suppression or alternation between the monocular channels of each eye. Instead, it is due to alternations between the stimulus itself.

Now Logothetis was ready to look for where in the brain this major event is going on. He found it in the ventral stream, way down in the temporal lobe. Cells in this part of the visual system are wholly dedicated to perception. Briefly, in the part of the brain that processes information about objects, there is a high correlation between neuron activity and which stimulus is being perceived. The brain automatically carries out this switching, this updating of the nervous system about a strange situation. Each eye sees something different.

. . .

Nowhere is the automatic brain giving us more done deals than in the emotional system. The feelings we have about the things we do are always cropping up and biting us. Almost everything we see, do, and hear evokes a salience within that reflects subtle but continual conditioning, which proceeds outside the realm of awareness.

Antonio and Hanna Damasio, two of the most inventive people in brain research, scientists who always provide new ways of thinking about difficult problems, have been hammering away at understanding how emotions contribute to our cognitive lives. The Damasios have postulated that we choose cognitive strategies because our viscera signal our brain about which ideas we should use in any given situation. Day-to-day and minute-to-minute decisions require not only cognitive planning, but also interactions with our past. That history has an emotional component shaped by success or failure in what we did or what occurred.

You need a measurement standard when trying to clarify such complicated things. The Damasios and their colleagues use a simple card game to illuminate how emotions interact with cognition automatically and outside of awareness. Four decks of cards and a pile of money are presented to a player. The task is to win as much money as possible. Players start turning over the cards from each deck. Each card in decks A and B gives an immediate $100 reward; on decks C and D the immediate reward is only $50. It is easy to figure out that turning over the cards in decks A and B will generate more money, but the cards have been arranged so that unpredictably large losses can also occur—and more often with decks A and B than with C and D. Players have no way of predicting when a penalty will be given, no way of accurately figuring the net gain or loss from each deck, and no knowledge of when the experimenter will terminate the game. Normal subjects usually reason things through and begin to pick from only decks C and D.

The Damasios discovered two major things with this test. They had hooked up skin electrodes to normal subjects so they could measure their galvanic skin response (GSR). We all sweat a bit when we are emotionally tied up in an event. The sweating changes the electrical conductance capacity of our skin, which can be detected with a recording device. The Damasios noted something astounding. Before subjects had figured out the game and begun to concentrate their responses on decks C and D, the skin seemed to know to do this! That is, a GSR response occurred for decks A and B, and eventually this response cued the subject to avoid these decks. But the response happened before subjects could explain why they were picking up decks C and D.

The second discovery was that patients with lesions in their prefrontal lobe perform quite differently on this task. Those with prefrontal lesions of the ventromedial area suffer from an inability to make real-life decisions. They lose their jobs, go bankrupt, and get divorced. They can't seem to calculate the difference between short-term and long-term gain and loss. It isn't surprising, then, that in this test such patients never learn which decks will yield a long-term gain.

Our automatic brain files away gain and loss experiences, and when we encounter a new decision, the emotional brain helps us sort out which cognitive strategy to use, even though for a startlingly long time we remain unaware of why we are doing one thing rather than another. Prefrontal patients can't take advantage of this unconscious cuing because, as the Damasios believe, the lesion interrupts the pathways for critical information arising in the gut to be communicated to the decision processes in the brain. Now hold your hat.

A few months back a young French doctor who had just completed his thesis and was now visiting at Yale walked into my office. He was attending a meeting on biology and ethics at Dartmouth and wanted to talk about Immanuel Kant's brain lesion. His *what*? Dr. Jean-Christophe

Marchand had been reading about Kant's life and medical history. Until Kant reached the age of forty-seven or so, his writings are straightforward and, believe it or not, clear. After this age Kant began to write his great philosophical works, which emphasize the idea that innate cognitive structures exist independent of emotions. Nearly impossible to read, his works make Jean Piaget's "writing" seem lucid. But Marchand's points are tantalizing. Kant began to complain of headaches and other maladies and gradually lost vision in his left eye. Dr. Marchand deduced that Kant had a left prefrontal lobe tumor—growing slowly, but there. Damage to this area affects language ability and the ability of our emotional system to cue us toward good cognitive strategies. Is it possible that all those Kantians have saluted a man who was writing nonsense—a philosophy for those who do not have a normal cognitive and emotional system?

. . .

Our conscious lives depend on all kinds of automatic processes happening inside our brains. Though we can't even influence them by willed action, we continue to believe that we are in control of what we do. As we move away from perceptual processes and delve into more cognitive issues like memory, we find an even greater separation between the actions of our automatic brains and our spin-doctoring minds.

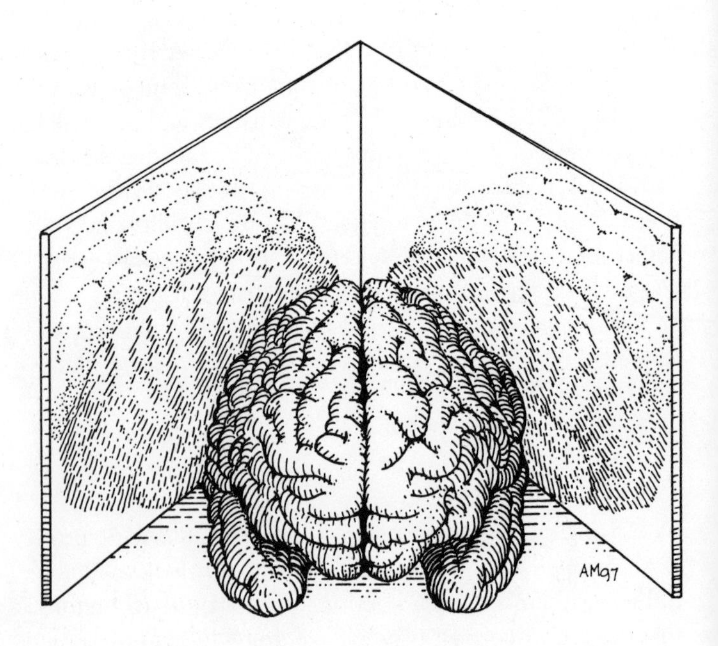
AM97

6 REAL MEMORIES, PHONY MEMORIES

The next stage is memory, which is like a great field or a spacious palace, a storehouse for countless images of all kinds which are conveyed to it by the senses. In it are stored away all the thoughts by which we enlarge upon or diminish or modify in any way the perceptions at which we arrive through the senses, and it also contains anything else that has been entrusted to it for safe keeping, until such time as these things are swallowed up and buried in forgetfulness. When I use my memory, I ask it to produce whatever it is that I wish to remember.

ST. AUGUSTINE, *Confessions*, BOOK X

If only it were true. The tidy thought that our memories accurately reflect our past is a pervasive one, for it leads to the view that the mind calls on an orderly warehouse full of neatly filed packets of memory. We can readily summon these memories, whether they be a telephone number from childhood or an image of a mother's face; prima facie, memory works remarkably well. Lots of neuroscientists want the memory system to work this way, too. So do many clinical psychologists and trial lawyers who think that memories are accurate representations of the past.

Sometimes memories are repressed—a door is temporarily closed to one of the orderly warehouse bins. But when it opens, anything that comes out of the mouth is to be believed as an accurate representation of a past event. Many don't want to believe that retrieved memories may be false or confused. Even laboratory brain scientists sometimes forget what is known about how memory works at the psychological level and claim there are orderly brain processes. After all, they want to locate the phenomenon, analyze, manipulate, disrupt, and improve on it, and do what they do. It is much easier to understand a static and real entity than a fuzzy, moving one.

Even so, our personal memories don't enjoy such a status in our brain and mind. We discover this in the normal course of events as we shuffle toward our gray decades. The annoyance of forgetting proper names comes first as a warning flag that things are not what they were. Remembering a story but not who told it to us is another intrusion on our orderly mind. The number of options we can consider while we decide what to do becomes fewer. And our recall of events stinks, even though we're better at recognizing items. I can remember the time I tried to invite Richard Dawkins to a meeting in Napa Valley. I flattered him, bribed him, tried everything. I made only one mistake. In my letter I addressed him as "Dear Dr. Hawkins." Oh well.

All these well-known aspects of our aging memory system are accepted even though everyone tries to paper them over. They are real and they are there. But you may be surprised to learn they are there in spades in young brains, too. Starting in childhood and going up to our forties, our brains are cooking up untrue stories about our past. They can't help it. It's due to the way our brains are organized for memory.

We know little about how memories are stored in the brain. The topic so naturally appeals to us that our enthusiasm for knowing more about it frequently outruns our rational powers to carefully think about any new claim. Back in the fifties the distinguished neurosurgeon Wilder Penfield and his colleagues at the Montreal Neurological Institute were busy stimulating the temporal lobes of human patients. Their epileptic patients were undergoing surgery to remove diseased areas in the brain that give rise to epilepsy. Penfield stimulated their brains while the patients were under local anesthesia. A few of them would suddenly recall a memory moments after being stimulated, and Penfield claimed this shows that memories are stored in highly localizable areas of the brain. This idea was taken up and to this day lurks around in one form or another. But no one at the time checked on the veracity of these memories. Now we know that the details reported in those episodes had little basis in reality. The work of Elizabeth and Jeffrey Loftus at Washington University and Larry Squire at the University of California, San Diego, has led us to believe that the probings activated some kind of generic memory of the past or a full-blown false memory. This idea conflicts with those of the famous American psychologist Karl Lashley of Harvard University. He kept training rats to do all sorts of things, then produced lesions in their brains in hopes of finding the engrams or specific memories associated with one of his trained tricks. He searched and searched and never found a brain area that represented a particular memory. He concluded that memories are distributed throughout the brain in some fashion.

The current view, that a memory really reflects a constellation of functions that are networked together, is a hybrid of these two earlier ideas. Each aspect of the constellation is most likely managed by a local area in the brain. So now we talk about network specificity rather than regional specificity when it comes to describing how memories are stored in the brain.

Daniel Schacter at Harvard University, who has recently been working tirelessly on the issue of false memories, deals articulately with this issue of network specificity. He has proposed a model, called the Constructive Memory Framework (CMF), which is an integrative framework for several processes in the memory system. CMF must solve a number of problems, during either encoding or retrieval, in order to produce an accurate memory. For example, we can falsely recall an event if we fail to bind together the separate elements of an experience during the time of encoding. Each of our experiences takes place in time and space and in an affective state. They are all part of the memory in question. If one of the many elements associated with a past memory is called up incorrectly, the resulting memory may well become distorted before our very eyes.

An occurrence in my family nicely illustrates CMF. I am not a bad cook, and over the years I have perfected my spaghetti alla carbonara. Marcela Hazan got me going with my favorite recipe, but then it was uniquely modified by the maître d' at New York's Piccolo Mondo restaurant on the Upper East Side. One night after listening to my recipe he told me his is much better but that it was not on the menu. He agreed to make it nonetheless and it was divine. The trick is to use consommé instead of cream when preparing the pancetta. It makes it all so light and wonderful.

When my sister-in-law came to visit, she went absolutely wild for the dish. She had never heard of such a thing in Texas and was determined to add it to the family repertoire. She did and we were all eager to taste it. Her creation, too, was divine, but it had nothing to do with my recipe. She misremembered virtually every ingredient except the pancetta! She swears the two recipes are the same, comparing hers to ours in her mind. We all do such things all the time, combining elements of an event with misremembered elements and coming up with unique experiences.

Such everyday stories about how our memory systems work underline the notion of a network that can store information in the brain, and this basic fact is instructive on many levels. First it points up that one should never look for a single event in life to be represented in one neuronal cell, although individual neuronal cells must somehow configure themselves to alter their firing pattern in a way that influences the network. Scientists are working on these issues and coming up with exciting new developments. Using molecular genetics techniques, for example, Susumu Tonegawa at MIT and Eric Kandel at Columbia University selectively damaged a cell system in a brain area known to participate in memory, the hippocampus. The injury had a marked effect on the animal's spatial memory capacity, but none on the animal's ability to learn other kinds of problems. Their enormous achievement is in transiently knocking out, or turning off, genes which control the development of an organism. Instead of totally blocking gene actions—a technique that may be too disruptive for analyzing neuronal cell function—molecular biologists can now turn off the gene in a more selective way by disabling it for short periods during development.

In moving on from the neurobiology of memory, I am compelled to note that, once again, automatic processes are at work in the brain. Stuff gets stored effortlessly and explicitly, as when we learn a new language, or implicitly and incidentally, as when we play tennis and gradually improve even though we don't know why. In both instances the brain stores things in some marvelous and automatic way.

The fun begins when we turn to thinking about all the ways of getting information *out* of the brain; these range from wildly inventive to flat-footed to mundane. As we grow aware of how closely our memory is tied to other mental processes, our sense of our mental life's intricate and interarticulated nature becomes even more wondrous. Take, for example, the extraordinary tricks magicians play on our mind and memory. They use the simple device of redirecting our attention to make objects that are in our full view, that we know our retina transmitted into our brain, go unnoticed.

Harry Blackstone Sr., a truly great magician, could pull off a trick relevant to this condition without being outguessed even by fellow magicians. He gave the trick major billing on his posters, calling it "Out of Your Hat." Standing beside a large top hat, Blackstone would tip it up and show the audience there was nothing under it, replace it, switch his attention and that of the audience to his beautiful assistant standing on his right, pull parts of a sheet out of a container, and hand them to his assistant, one wad after another. After a minute or so he would pick up a sheet with great gusto, turn to his left, and throw it down over the large hat that had been there all the while. The crowd was ready. What could be the trick? Blackstone then whipped the sheet and hat back, and there was a full-sized donkey—not a rabbit or a chicken, a damn big donkey. Unbelievable.

How did Blackstone do it? Very simple. While he was directing the audience's attention to the beautiful assistant and keeping them busy with the sheet exchange, another assistant walked on stage with the donkey in tow and placed it next to Blackstone. When the second assistant walked off the stage, Blackstone, with the deftness and alacrity of a great matador, whipped the sheet over the top hat. The stage was now set for the trick. The donkey and the second assistant carried out their act in full view of the audience, which was paying attention to Blackstone. The fact that the second assistant walked the donkey out in full view and placed the hat on top of it was transmitted to every brain in the audience, but the brains didn't register the information. Truly amazing! Next time you go to a magic show, even though you'll be armed with this now-public secret, you probably won't be able to detect a trick like this.

The talented psychologist Don Simons of Cornell University has delved into the vast importance of attention in memory research. Simons has videotaped wonderful examples of people not noticing huge changes in their personal environment if they are otherwise engaged in an activity. One memorable example is two undergraduates, one of them a stooge for the experimenter, talking about something on a lawn. In the middle of their conversation, a workman, carrying a large door, walks between the two. During the moment the two are blocked from one another's vision a second stooge replaces the first. After the conversation winds down, the unsuspecting student is asked about the event and whether he noticed anything odd. The student emphatically swears nothing unusual happened, even though the person with whom he started the conversation was not the person with whom he finished it! From the beginning of any experience the automatic brain toils away beyond our control. Many factors influence what becomes part of our personal memories or what we think are our memories.

. . .

Of all the syndromes in neurology and of all the discoveries in brain science, none is more wondrous than the behavior of a split-brain human patient. I give myself license to make this debatable assertion because I have studied these patients for almost forty years. Every time they are examined, they reveal truths about how brain enables mind.

Epilepsy strikes in many ways and for many reasons. Sometimes it grows out of birth trauma or subsequent trauma. Sometimes it is metabolically induced or is a side effect of a tumor. Sometimes the origin is a complete mystery. However it starts, it is terrifying. Imagine not knowing when you might without warning lose consciousness or drop to the floor in a convulsion. Imagine the psychological and social toll not only on you, but also on those around you. Epilepsy is serious stuff, and those who have given of themselves to remedy this Dostoyevskian disease have helped humanity in a big way.

The disease can usually be controlled with medications that penetrate the blood-brain barrier to neutralize the cerebral tissue that triggers a seizure by adjusting the seizure site's chemistry. But many patients cannot be helped by medications alone; these are the ones cared for by neurosurgeons. They have two methods at their disposal: remove the epileptic tissue or split the brain. Both have been exceedingly helpful. When a known seizure site or focus is localized to a part of the brain that can be spared, it is generally excised. But multiple foci render excision impossible and supply a strong rationale for splitting the brain. Disconnecting the two hemispheres localizes a seizure to the hemisphere in which it begins. During the seizure the other half-brain remains in control of the body. The patient stays conscious and in charge during the attack.

The operation was first performed in modern times by Joseph Bogen and Peter Vogel at the Loma Linda Medical School in California. It fell to the late Roger Sperry of the California Institute of Technology and me to examine the consequences of this surgery on behavior and cognition. It was a thrilling time.

First and foremost the surgery works. Seizures are controlled and patients return to a normal life. The medical end of the story is rich and fulfilling. Split-brain effects have to be exposed in a laboratory, where special techniques separately test each half-brain. From such techniques we can discover the amazing effects of disconnection. My interests have always been directed at what disconnecting the two cerebral hemispheres can teach us about ourselves. Over the years hundreds of experiments have been carried out, and they mainly reveal that the thoughts and perceptions of one hemisphere go on outside the realm of awareness of the other. The left brain is crammed with devices that give humans an edge in the animal kingdom. This is the hemisphere that is adept at problem solving and thinking. While the right brain is better at things like facial recognition, the left brain is crucial for our intelligence agency. The split-brain patient appears to have two minds. What the left brain learns and thinks is unknown to the right brain, and vice versa. This dramatic result is a centerpiece of brain research.

What we learned after doing the research for about fifteen years is that the right hemisphere can be instructed to do something or to react to an emotion and the right hemisphere will respond to the specific command we give. Even though time and again the left brain could not tell us about the command given to the right brain, it didn't seem perturbed about the right brain carrying out whatever the command might be. There was never a complaint about this odd state of affairs.

It was on a snowy trip to New England when my colleague Joseph LeDoux and I came up with an intriguing idea: ask patients (which is to say, ask their left-speaking hemisphere) how they feel about the right hemisphere occasionally doing things outside the left hemisphere's realm of control. Relying on the experimental approach I describe in Chapter 1 where we showed a picture of a chicken to the left brain and a snow scene to the right, we set to work and discovered that the left brain has a specialized mechanism that interprets actions and feelings generated by systems located throughout the brain.

Reenter the interpreter, the means of analyzing things such as why a feeling changes and what a certain behavior means. For example, give a command to the silent, speechless right hemisphere: "Take a walk." Then see how a subject typically pushes back her chair from the testing table and starts to walk away. You ask, "Why are you doing that?" The subject replies, "Oh, I need to get a drink." The left brain really doesn't know why it finds the body leaving the room. When asked, it cooks up an explanation.

There it was. The half-brain, which asks how A relates to B and constantly does that when solving problems, is also the hemisphere that provides our personal narrative for why we feel and do the things we feel and do. Even though, as in the split-brain patient, feelings and actions are precipitated by a brain system operating outside the left brain's realm of knowledge, the left brain provides the string that ties events together and makes actions or moods appear to be directed, meaningful, and purposeful.

Once LeDoux and I grasped the power and centrality of the interpreter, we saw it at work everywhere. Take the panic attack patient. For no apparent reason a sudden feeling of panic comes over otherwise normal people. It is an awful feeling, one of free-floating, intense anxiety. People who have these attacks can be helped medically; indeed, medicines can be so effective that they never again have an attack. But psychiatrists have noted that the dozens of phobias these people possess do not go away so easily. Over the course of their disease they could readily cook up a theory for why they had the attack—they were with people, in a certain restaurant, or with a particular person. They interpreted these things and decided they didn't want any part of the person or place again; they had developed a phobia. Curing the onset of the panic doesn't instantly change the memories of these earlier interpretations as to why the attack happened. Psychiatrists have come to realize that good old psychotherapy can reverse these earlier interpretations, these phobias.

Neurology yields even weirder examples of how the interpreter works. Our understanding of these bizarre syndromes is heightened by knowing about the interpreter. Take for example a malady called *anosagnosia,* which means people deny awareness of a problem they have. Those who suffer from right parietal lesions and are hemiplegic and blind on the left side frequently deny they have a problem and claim the left half of their body is not theirs! They see their hand, but maintain it has nothing to do with them. How could that be?

Consider what happens to a lesion in a human's optic tract. If the lesion is in a nerve that sends information about vision to the visual cortex, the damaged neurons no longer work; hence a patient complains she is blind in part of her visual field. Another patient has a lesion in the visual cortex that creates a blind spot of the same size and in the same place, but he does not complain at all. He doesn't say anything is wrong because his cortical lesion is in the place in his brain that has the representation of that exact part of the visual world. It is the part that asks, "What's going on to the left of fixation?" It is functioning when a lesion occurs on the optic nerve; but now the nerve can't get any information and the cutoff brain puts up a squawk: "Something is wrong. There is no information coming in." When that same part of the cortex suffers a lesion, however, the patient's brain no longer cares about what is going on to the left of fixation; no squawk is raised because the system that generated the squawk is gone. In short, just as our brain doesn't complain because it can't see behind us—there is, I'm sad to say, no part of the brain that can do it—the patient with the central lesion doesn't have a complaint because the part of the brain that might complain has been injured.

As you move farther into the brain's processing centers, this same condition applies. Now lesions that affect mental processes begin to interact with the interpretive function. The parietal cortex is where the brain represents how an arm is functioning. The parietal cortex constantly seeks information on the arm's whereabouts, its position in three-dimensional space, and monitors the existence of the arm in relation to everything else. If there is a lesion to sensory nerves that bring to the brain information about where the arm is, what's in its hand, whether it is in pain or feels hot or cold, the brain communicates that something is wrong: I am not getting input. But if the lesion is in the parietal cortex, that monitoring function is gone and no squawk is raised though the squawker is damaged.

A patient with a right parietal lesion has suffered damage to the area of the brain that represents the left half of the body. It cannot feel the state of the left hand. When a neurologist holds this patient's left hand up to his face, the patient gives a very reasonable response: "That's not my hand, pal." His interpreter, which is intact and working, can't get news from the parietal lobe because the lesion has disrupted the flow of information. The left hand simply doesn't exist anymore, just as seeing behind the head is not something the interpreter is supposed to worry about. So the hand being held in front of the patient can't be his.

An even more fascinating syndrome is one called *reduplicative paraamnesia*. I once studied a patient with this syndrome, an intelligent lady who, although she was being examined in my office at New York Hospital, claimed we were in her home in Freeport, Maine. The standard interpretation of this syndrome is that the patient has made a duplicate copy of a place (or person) and insists that there are two.

I started with the "so where are you?" question. "I am in Freeport, Maine. I know you don't believe it. Dr. Posner told me this morning when he came to see me that I was in Memorial Sloan Kettering Hospital and that when the residents come on rounds to say that to them. Well, that is fine, but I know I am in my house on Main Street in Freeport, Maine!" I asked, "Well, if you are in Freeport and in your house, how come there are elevators outside the door here?" The grand lady peered at me and calmly responded, "Doctor, do you know how much it cost me to have those put in?"

This patient has a perfectly fine interpreter working

away trying to make sense out of what she knows and feels and does. Because of her lesion, the part of the brain that represents locality is overactive and sending out an erroneous message about her location. The interpreter is only as good as the information it receives, and in this instance it is getting a wacky piece of information. Yet the interpreter still has to field questions and make sense out of other information coming in that is self-evident. As a result, it creates a lot of imaginative stories.

Before filing this vignette away as an oddity, think about how nutty a normal person can be in trying to put together two disparate facts. Just the other day I was checking out a four-wheel-drive pickup truck. My kids were in the backseat, and I was test driving it with the salesman riding shotgun. We were cruising along when I asked if I could shift the car into four-wheel-drive. "Sure," smiled the ever-agreeable salesman. After wrestling with the gears, I slipped the transmission into what seemed to be the proper gear, and the truck immediately jerked, heaved, and bounced along the perfectly dry, paved road on a beautiful summer day. Something was obviously wrong, but not for the salesman who wanted to sell me that lemon. "Oh," he said, "the four-wheel drive doesn't work well on dry roads. It has to be icy out, and that's when you need the four-wheel drive." I heard my nine-year-old son and twelve-year-old daughter suppress their gagging on that one.

My all-time favorite has to do with the probably apocryphal tale that Kenneth Jandell, the talented expert on forgeries, relates as he describes the nutty justifications people give when being challenged about the authenticity of their documents. Someone claimed to have Hitler's skull, and an expert was called in to examine it. It was obviously not Hitler's skull; it was the skull of a young boy. When this was told to the forger, without missing a beat he retorted, "Oh yes, this is Hitler as a young man."

The interpreter tells us the lies we need to believe in order to remain in control. These seemingly extreme examples are really not so extreme. As the social psychologist Eliot Aronson put it, we all want to shift our beliefs so we can hang onto the proposition "I am nice and in control." This idea grew out of Leon Festinger's work on cognitive dissonance, which provides a huge insight into how we behave. Discovering the interpreter adds some brain science to the story, but the basic insight that humans have such a device comes out of Festinger's work.

It all started with a small grant from the Ford Foundation made to Festinger to study and integrate work in mass media and interpersonal communication. He and his colleagues took on the project, and to hear Festinger tell it, the seminal observation came from considering a 1934 report about an Indian earthquake. What puzzled them was that, after the earthquake, the vast majority of the rumors predicted an even worse earthquake. Why after such a horrendous event would people want to provoke further anxiety? Festinger and his colleagues concluded this was a coping mechanism that the Indian people had developed to deal with their present anxiety. In other words, since the earthquake had filled the population with grief, they had formulated an even greater future tragedy, compared to which the present state of things didn't look so bad. From this basic observation the theory of cognitive dissonance was born.

Festinger carried out one of his early experiments with his two close friends, Stanley Schachter and Lew Riecken, in Lake City, Minnesota, where a group of people had come to believe the prophecy of Mrs. Marian Keech. Months before the crucial day, the following headline and news report had appeared in the *Lake City Herald:*

> *Prophecy from Planet Clarion Call to City: Flee that Flood. It'll Swamp Us on Dec. 21, Outer Space Tells Subordinate.*
> Lake City will be destroyed by a flood from Great Lake just before dawn, Dec. 21, according to a suburban housewife. Mrs. Marian Keech, of 847 West School Street, says the prophecy is not her own. It is the purport of many messages she has received by automatic writing, she says. . . . The messages, according to Mrs. Keech, are sent to her by superior beings from a planet called "Clarion." These beings have been visiting the earth, she says, in what we call flying saucers. During their visits, she says they have observed fault lines in the earth's crust that foretoken the deluge. Mrs. Keech reports she was told the flood will spread to form an inland sea stretching from the Arctic Circle to the Gulf of Mexico. At the same time, she says a cataclysm will submerge the West Coast from Seattle, Washington, to Chile in South America.

Your ordinary scientist might have stayed as far away from this as possible. This is *National Enquirer* stuff and potentially hazardous to one's career. But Leon and a team set out for Lake City, where Mrs. Keech received another message on December 20. An extraterrestrial visitor was to appear at her house around midnight to escort her and her followers to a parked flying saucer and take them away from the flood, presumably to outer space.

Festinger's prediction, assuming the momentous event did not occur, was that the followers would attempt to reduce their dissonant state at having their beliefs disconfirmed by attempting to convince others of those beliefs. There is now a vast set of experimental data to support that view, but at the time it was brand new. In Lake City that evening, as the clock struck twelve and no alien visitor arrived to take them to the spaceship, an awkward period began among the believers waiting in Mrs. Keech's living room. But a few hours later Mrs. Keech received another message:

> For this day it is established that there is but one God of Earth and He is in our midst, and from his hand thou hast written these words. And mighty is the word of God—and by his word have ye been saved—for from the mouth of death have ye been delivered and at no time has there been such a force loosed upon the Earth. Not since the beginning of time upon this Earth has there been such a force of Good and light as now floods this room and that which has been loosed within this room now floods the entire Earth. As thy God has spoken through the two who sit within these walls has he manifested that which he has given thee to do.

Suddenly, everyone in the room was in better shape, and Mrs. Keech reached for the phone to call the press. She had never done this before, but now she felt that she must, and soon all the members of the group had called various branches of the news media. This sort of justification went on for days, confirming Festinger's prediction in a spellbinding way.

These stories of blatant fabrication are funny but true. Everyday examples can be heard by listening to the interpreter in young children as they invent wild explanations of how the world works—as expected from a limited knowledge at that time in development. Or watch the nightly news analysis of why the stock market did whatever it did. The interpreter never rests. It is always trying to make sense out of what is going on around us. And it also reconstructs our past.

. . .

I like false memories. Some of my best memories are false, I am sure. That wonderful carbonara my sister-in-law now cooks is built up from a false memory. We all have an abundance of them, which is why Malamud's lament that I quote in Chapter 1 is so meaningful. Over the past few years the topic has taken on mythic proportions as trial lawyers and clinical psychologists have discovered a gold mine of reasons why Jones or Smith suffers from this or that. Think of all those repressed memories feeding the unconscious, which, in turn, created mental problems. As psychologists dig for these repressed memories, they discover god-awful things that happened at an early age. It is astounding. All of a sudden someone's daughter, in the hands of the so-called health professional, has realized that her father was a creep and molested her. This ugly story is passed on, unaffected by facts or what we now know about memories and how they go awry.

Remember John Dean, Nixon's White House lawyer? People credited him with having a photographic memory because of the detailed and vivid account of his conversations with Nixon in the Oval Office he gave to the Watergate committee. However, once the tapes of these conversations were released, it was discovered that, although he got the general idea of them correct, his memory for the details of the conversations was full of errors.

Our brains are built to remember the gist of things, not the details. It is as if our memory system can get us into the arena of a past event, but it can't recall the details of the event with much accuracy. Yet we frequently go on and on about the details as if we truly remember them.

Owen Flanagan, a very clever philosopher at Duke University, recounts how he once horrified his parents by relating to them how much he missed his boyhood friend, Harry. Flanagan had never had such a friend. He had mistakenly confused things in his past, and out popped Harry, whom he thought about all the time. Years later, when he was confronted with the truth—that Harry never existed—Flanagan pointed out that the fact made little difference to him. He had drawn plenty of wisdom and solace from Harry over the years, and Harry's steadfast friendship had served him well in his development. We all have Harrys—either in whole or in part.

The interpreter and the memory system meet in these false recollections. As we spin our tale, calling on the large items for the schema of our past memory, we simply drag into the account likely details that could have been part of the experience. Such easily disprovable fabrications go on all the time.

A way to find memory distortions is to present someone with pictures of a likely event. Show a group of normal people forty slides depicting a person arising in the morning, getting dressed, having breakfast, reading the paper, and going to work. After the subjects view this pictorial story, send them off for a couple of hours and then bring them back to the laboratory for testing. In a typical test some of the exact same pictures are mixed in with other pictures that could have been part of story; also included are pictures that have nothing to do with the story. All the subjects have to judge is which pictures they saw before. Not surprisingly, they make all kinds of mistakes, but mistakes of a certain type. They swear they saw many of the related pictures, and they are quite good at rejecting the unrelated ones. In short, they remember the gist of the story, and with that comes the schema of what logically fits the vignette. This sort of memory invention is false memory at work.

Michael Miller at Dartmouth College has devised a simpler version of this test. He asks subjects to study pictures of strongly thematic scenes, for example, a stereotypical beach scene with lots of activity. Later, when he asks subjects if they remember seeing a beach ball, they likely will say yes even though there was no beach ball. Furthermore, when Miller probes subjects for their specific recollections of the beach ball, they go into a very elaborate description, such as seeing boys tossing a beach ball with red and green stripes.

Other research, which was orginally introduced by James Deese at Johns Hopkins University in 1959 and recently brought back by Henry Roediger and Kathleen

McDermott, points to the same conclusion. If subjects are given a list of words to study, such as "bed," "rest," "awake," "pillow," and so forth, they often recall hearing the word "sleep" as well. However, if normal English-speaking people are given Turkish words to learn and later are tested for their memory of them, they make few erroneous claims that related words had also been shown to them. They have no schema for Turkish, so the interpreter has nothing to fill in.

Once a phenomenon like this has been unearthed and can be reliably demonstrated and measured, it becomes a tool for studying how the mind works. In this instance one of the first follow-up questions was where in the process false memories arise. Is information erroneously encoded at the start of the experience, or is there a snafu as one tries to retrieve the information? Are the same brain mechanisms that are active in storing real memories active in storing false memories?

Miller and I recently tested at which stage of memory processing false memories occur using his schematic, picture paradigm and the associative word list paradigm just described. We had speculated that associative false memories are fundamentally different than schematic false memories and that they may be occurring during either encoding or retrieval, depending on the paradigm. For instance, the false memories created by the word lists may be happening during encoding, since the subjects are often thinking of the word "sleep" when they are studying the words "bed," "rest," "awake," and "pillow." During a test, whether subjects actually heard the word during the study session or thought of the word when they were trying to retrieve it could simply be a matter of confusion to them. With the schematic paradigm, however, subjects are unlikely to think of an item that is not there when viewing a complex picture, such as a beach scene. Only when they are asked if they remember a beach ball do subjects draw upon their preexisting schemas and create an image of the beach ball in the beach scene. In this case the false memory is being created during retrieval.

To test this idea we used various experimental manipulations to boost memory performance that are known to work during encoding. If a false memory is being created during encoding, it should benefit from these encoding manipulations. Indeed, the tests confirmed our speculations in that there were significantly more false memories from the word lists following the manipulations, whereas the manipulations had no effect on the false memories from the schematic paradigm.

Elizabeth Phelps at Yale University tested the question of which brain mechanisms are responsible for false memories on the left and right hemispheres of split-brain patients. Using a series of pictures depicting a story, she learned that the left brain, not the right brain, claims that some of the related pictures have been presented before. The interpreter-charged left brain remembers the gist of the story line and fills in the details by using logic, not real memories. The right brain, without an interpreter, regurgitates the literal story, not one embellished by the interpreter.

Apparently the problem with false memories comes from the interpreter. Indeed, Miller and I made images of the brain that show the left hemisphere's activities during tests that discover a subject recounting false memories. In using associative word lists, Miller observed that although both hemispheres are activated when recalling the true items, the left hemisphere becomes more activated when it is experiencing false reports.

Daniel Schacter has recently been studying the relationship between the production of false memories and damage to specific regions of the brain. In a group of patients with damaged medial temporal lobes who suffered from poor memories he found they actually produced fewer false memories than normal subjects. Of course, they produced fewer true memories as well. Schacter concluded that these patients not only had a deficit in remembering events from the past, but they also had deficits in remembering the gist of events, which can lead to false memories. Even more fascinating is his case of a patient with damage to the right frontal lobe. This patient, with an intact left hemisphere, actually produced more false memories than normal subjects. He was overreliant on the general characteristics of the event and was a slave to his left-brain interpreter.

Even cognizant of the powerful role the interpreter plays in creating our reality and coloring our experiences, one must be astounded by the groundbreaking science of Steve Ceci at Cornell University. In pursuit of the truth about the repressed memories of childhood, Ceci wants to know how normal children behave in response to suggestions provided by others. The fear is that when psychotherapists interview troubled children, they inadvertently implant in them the idea that some untoward event happened to them in their earlier years. Once the suggestion is made, can a child come to believe it even though the event never occurred?

Ceci and his colleagues uncovered startling insights to a child's mind in these circumstances. An examiner would show a preschool child a set of cards, each describing an event; some had really happened to the child and some hadn't. The child was asked to pick one of the cards, and the examiner read it. An example of a false event might be: "Got finger caught in a mousetrap and had to go to the hospital to take the trap off." Then, at various times over a ten-week period, the examiner asked the child, "Think real hard, and tell me if this ever happened to you. Can you remember going to the hospital with a mousetrap on your finger?" If the child answered affirmatively, the examiner asked follow-up questions about how the child got to the hospital and so on.

Although Ceci was willing to believe that false stories have an impact, he was not prepared for the astonishing discovery that 58 percent of the children tested claimed that at least one of the false events had happened to them and 25 percent produced false narratives for the majority of them. Moreover, the stories became more elaborate over time. While the first week or two after the exposure to the story might find the child simply saying the event had occurred, by ten weeks the stories had become full blown with rich details. One example went like this: "My brother Colin was trying to get Blowtorch [an action figure] from me, and I wouldn't let him take it from me, so he pushed me into the wood pile where the mousetrap was. And then my finger got caught in it. And then we went to the hospital, and my mommy, daddy, and Colin drove me there, to the hospital in our van, because it was far away. And the doctor put a bandage on this finger [indicating]." Although none of this happened, by the time the child told this full version, he was totally convincing. He made all the right gestures; he was emphatic about it; he was believable. He was ready for the courtroom.

. . .

Nowhere is our automatic brain in more trouble than in recalling the past. The interpreter, working from noisy data, compounds the problem by embellishing on what it does recall. The story remembered on one day becomes part of the memory for the next time it is told. Soon begins a rich narrative about past events. The narrative most likely becomes less accurate and much more elaborate in its detail. The old adage that so and so just can't see it, can't see he has certain negative features in his personality, is true. He has weaved another tale about himself. My bet is it works even under horrendous conditions. I bet O. J. Simpson thinks he is a good person.

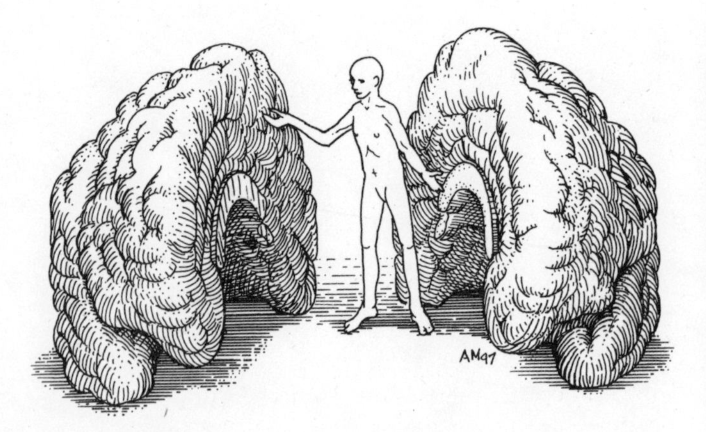
AM97

7 THE VALUE OF INTERPRETING THE PAST

Something there is that doesn't love a wall.

ROBERT FROST, "Mending Wall"

Just as humanists like John Updike have their thread that "runs through all things," as I say in Chapter 1, so, too, is the interpreter omnipresent in our lives. It toils away at duties from perception to memory. In general the interpreter seeks to understand the world. In doing so it creates the illusion that we are in control of all our actions and reasoning. We become the center of a sphere of action so large it has no walls.

The manifest presence of the interpreter, rearing its magnificent head above the sea of species around us, raises the question "Why us?" Is it really a special device, or is it the mere consequence of the brain getting too big and loaded up with neurons? Is it truly a human instinct, an adaptation that supplies a competitive edge in enhancing reproductive success? I think it is, and my guess is that the very device which emerged to help us conquer the vicissitudes of the environment enabled us to become psychologically interesting to ourselves as a species.

It all started long ago. Somehow during that still mysterious time of neuronal rearrangement an ancestor realized its brain could carry out simple reasoning. This new device might have enabled the brain to ask a question. Perhaps the brain noted nausea, and our ancestor was able to ask, "Why am I feeling so nauseated? Was it that smelly caveman we had over last night, or was it that weird green plant we ate, or what about that rotting meat?"

All kinds of animals can figure out what makes them sick. Rats can remember back to what they ate hours before and quickly learn to avoid food that makes them ill. That sort of simple associative learning abounds in the animal kingdom. What does not abound is the capacity to ask the next questions. "Why did the plant make me sick? How can I keep this sort of thing from happening again?" Sustained syllogistic reasoning, the capacity to state a major premise, then a minor premise, followed by a deductive conclusion, is what our species alone can do.

There isn't a normal member of our species who can't carry on sustained rational thinking. In the search for what makes the human act more intelligently one should not ask why Jones is a physicist and Smith is a janitor. Those are slight differences that result from a multitude of factors. The more basic question is why both Jones and Smith know how to ask the second-order questions: X is associated with Y. But why is that true? Perhaps Jones can do a little more than Smith, can take a few more steps into a problem. But we all can do it, and we do it reflexively.

Charlotte Smylie, my wife and confidante, and I asked the hemisphere with an interpreter what it could do in simple problem-solving tasks. We wanted to know what it has that enables it to do the things it can do. We asked the same question of the hemisphere without an interpreter, the right hemisphere. We were trying to learn whether one has a cognitive component that the other lacks. We were cognizant that each half-brain of a split-brain patient has essentially the same number of neurons, the same brain structures at the level of gross anatomy. There seems to be no a priori reason why one structure should be able to do something the other cannot.

We discovered that the left brain is easily able to go beyond information given in a test, yet the right brain can't. If, for example, a simple picture of a pan is presented to one or the other of the hemispheres and a picture of water is shown to the same hemisphere, the left brain can easily point to a picture of a pan with boiling water in it, but the right hemisphere cannot. It can't make the simple inference that pan and water could go together to become a pan of boiling water.

PET scans on normal subjects also reveal that the left brain participates in simple reasoning. Vinod Goel at York University in Canada, using brain imaging, offered visual proof that during a task which requires a subject to perform syllogisms, the left hemisphere is more active at reasoning than is the right one. For example, if the deductive syllogism

Some officers are generals.
No privates are generals.
Some officers are not privates.

is given, the subject must declare whether or not it is valid. As the subject figures out the syllogism, the left brain appears active in solving the problem. This is also true when the problem is on spatial relations. That is, even when the problem seemingly calls upon the right hemisphere's unique skills for grasping spatial relations, the left still dominates when it comes to reasoning about those relations.

Goel reminds us of the ubiquitousness of syllogistic reasoning, which is essentially the act of reaching conclusions from a limited amount of information. He cites the most celebrated reasoner in the Western world, Sherlock Holmes, as he demonstrates his left-brain capacity to Dr. Watson in "A Scandal in Bohemia" by Sir Arthur Conan Doyle:

> Then he stood before the fire and looked me over in his singular introspective fashion.
>
> "Wedlock suits you," he [Holmes] remarked. ". . . And in practice again, I observe. You did not tell me that you intended to go into harness."
>
> "Then how do you know?"
>
> "I see it, I deduce it. How do I know that you have been getting yourself very wet lately, and that you have a most clumsy and careless servant girl? . . . It is simplicity itself," said he; "my eyes tell me that on the inside of your left shoe, just where the firelight strikes it, the leather is scored by six almost parallel cuts. Obviously they have been caused by someone who has very carelessly scraped round the edges of the sole in order to remove crusted mud from it. Hence, you see, my double deduction that you had been out in vile weather, and that you had a particularly malignant boot-slitting specimen of the London slavey. As to your practice, if a gentleman walks into my rooms smelling of iodoform, with

> a black mark of nitrate of silver upon his right forefinger, and a bulge on the right side of his top-hat to show where he has secreted his stethoscope, I must be dull, indeed, if I do not pronounce him to be an active member of the medical profession."

Mere mortals use simpler examples every hour of every day. Doing research takes a grant. I didn't get my grant. Therefore I won't be doing research. Or the Yiddish variant: "Harry, get up and close the window. It's cold outside." Harry gets up and closes the window, turns to his wife, and says, "So now it's warm outside?" These events are in our everyday realm, but even these reflexive thoughts are a product of only one side of the brain.

We have known for years that left-hemisphere damage is bad news for language and thought and right-hemisphere damage has its impact on spatial tasks. Indeed, people with lesions in the temporal-parietal lobe are terribly disrupted on language tasks, particularly the comprehension of language, while injury to more anterior regions finds patients disrupted in the production of language. Now there is a large cottage industry in neuropsychology that tries to figure out the details of the impact of left-brain lesions on the many subtle properties of language. It is a difficult task because lesions are never the same, the brain varies in local organization from one person to another, and linguists are not always in agreement about what a particular test means. But all concur that the left brain is the site for language and thought.

The variations in mental capacity observed following the erratic left-brain lesion are also remarkable. These are usually made by Mother Nature through stroke and can be quite irregular. In one patient an infarct of a particular artery can knock out one zone of cortex while sparing another. In the next case the opposite pattern can occur. The net effect can be incredible dissociation of functions within the left hemisphere. The varying lesions to this hot zone of mental function can produce a patient who is devastated with respect to language function, but relatively spared for the capacity to think and solve problems. And the converse can be true as well. In the latter case the heartbreak of Alzheimer's disease commonly finds the patient easily able to talk, but running on empty cognition.

So there is something very special about the left brain. It reflexively generates a notion about how things work even when there is no real event to figure out. There is no stopping it. If you want to see it in action for yourself, try out this old parlor game at your next dinner party. It is an amazing thing to see. Before you begin the stunt, write down a list of about thirty "yes"and "no" answers. Randomly mix them up, but end the list with four "yes" answers in a row. You are ready. Now say to some poor unsuspecting soul at your party: "Tell me about the rule I am thinking about concerning a number sequence I am thinking about between one and one hundred. When you are right, I will say 'yes,' and when you are guessing wrong, I will say 'no.'" It's particularly fun if a know-it-all takes the bait. He might say twenty-six, and you, looking at your totally random list of yes-no responses, answer "no." He says thirty-two and you respond "yes." This goes on for a while, you supplying the know-it-all with a totally random set of yes-no answers to his guesses. And then you respond with three or four "yes" responses in a row. You stop and tell the subject he is doing well and ask him for his theory.

Unbelievable as it may seem, everyone has one! The person guessing usually beams with his success, very proud of his accomplishment. Then you ask what his theory is. At this point a frightening spew of gibberish erupts. The left brain insists on generating a theory, even though there really is nothing to generate one about. The guest, usually mortified, will not talk to you for a month.

It is not so surprising that the guests try to figure out the answer. After all, they think you are giving them some kind of IQ test, and they have no reason to believe you are duping them. What is so surprising is their continuing and natural attempt to try to see order where there is none and the fact that after they propose a theory, they actually believe it. All day long we are giving running accounts of why things are working the way they are. By and large, when the information is relatively simple, we are usually right about our stories, and we are reinforced.

The left hemisphere's capacity for continual interpretation means it is always looking for order and reason, even when they don't exist. No one has demonstrated this more dramatically than George Wolford at Dartmouth College. In a test that requires them to bet if a light is going to appear at the top or bottom of a computer screen, humans perform in an inventive way. The experimenter manipulates the stimulus to appear on the top 80 percent of the time, but in a random way. While it quickly becomes evident that the top button is being illuminated more often, subjects keep trying to figure out the whole sequence and deeply believe they can. Yet by adopting this strategy, they are rewarded only about 68 percent of the time. If they always press the top button, they are rewarded 80 percent of the time. Rats and other animals are more likely to learn to maximize and press only the top button. It turns out that our right hemisphere behaves like a rat's. It does not try to interpret its experience and find deeper meaning. It continues to live only in the thin moment of the present. And the left, when asked to explain why it is attempting to figure out the whole sequence, always comes up with a theory, no matter how spurious.

In most cases it would seem perfectly rational that the left brain should look for order. If there is order for which the left brain failed to look, it would lose the ability to make predictions. Thus there can be great payoff for it always looking for order, and we can't fault the left brain for looking when there is none. It doesn't really know there is none; only the experimenter does in these tests.

We humans have met our limit. Our interpreter works beautifully to help us understand the world. It fails us when trying to interpret giant data sets or meaningless data sets. And yet in that failure we gain a comfort that we think we know, even though we don't. We see connections where there are none. Perhaps this is why hope springs eternal!

. . .

The first step up from animal cognition, the capacity to go beyond the information given, served our species well for hundreds of thousands of years. The brain machinery present in our left hemisphere that allows us to ask how X is related to Y might have caused early humans to wonder about food caches, the neighbors, and the weather. In a simple world it obviously would have reaped huge advantages. But sustained rational thought is another matter, and mathematical thinking is even less omnipresent.

Leon Festinger, the late and great social psychologist, was hands down the most rational man I ever knew. In his book, *The Human Legacy,* he marvels at the inventive human mind, but then points out how little it takes to be inventive and how bad we are at sustained rational thought and mathematical reasoning. Just ask around and determine how many people you know who remember what a log unit is in mathematics. Or ask how you convert a number from base seven to base three. Ask them what that even means.

Mathematical thinking and statistical reasoning have been the subject of a huge amount of research over the past thirty years. Herbert Simon, the well-known Nobel laureate at Carnegie Mellon University, coined the term *bounded rationality,* which refers to when human reasoning abilities break down and this starts to create problems for how humans come to think about problems. Finding examples of how often and how quickly we break down is dead easy. As a species our intuitive statistics seem to be wanting.

Daniel Kahneman and the late Amos Tversky explored this issue with countless examples of how intuitive reasoning powers break down and create erroneous conclusions. They were looking for what natural reasoning powers we bring to an inductive reasoning problem. In one of their classic examples, they asked whether days with over 60 percent of male births were more common in a hospital that had forty-five births per day than in one having only fifteen births. Or would male births be equally common in both hospitals? What do you think?

If you're like me, you would say it doesn't matter—and you would be wrong. In responding with hasty rational judgments, we confuse a heuristic with a law: what Kahneman and Tversky call the *representativeness heuristic* and the law of small numbers. In a quick moment we fail to realize that the law of small numbers holds that the larger a sample, the less likely it will deviate from 50 percent for a random variable like the sex of a newborn. We would be more likely to get a 60 percent figure in a smaller sample. That statistical fact plays second fiddle to the heuristic that both samples, forty-five and fifteen, seem good enough to be equal to the perfect sample size that would ensure a 50 percent ratio.

Kahneman and Tversky are after rules for how the mind operates when making a decision. While some see this twosome as arguing that humans are, at their core, irrational, they really are saying something quite different. They are maintaining that even though people in the foregoing example are operating from an incorrect assumption, they still proceed to think rationally.

There is nothing more amusing than to sit in on a faculty meeting of psychologists, people trained in statistics, and to hear them go on and on about the disastrous acceptance rate of graduate students that year. Normally a department might add on two or three a year, but this year none chose to accept the offer. Does this mean the reputation of the department is slipping? After great handwringing and teeth-gnashing and finger-pointing, someone mentions the law of small numbers. Everyone pauses for a minute, but quickly jumps back into the fray.

The failure to appreciate basic statistics is rampant. Nowhere is it more telling than in the area of drug testing. Wolford examined this issue. School after school takes part in drug testing, and none of the administrators is thinking with any clarity. Yet the reason drug testing is even being considered is that the average Joe has trouble grappling with the base rates of things happening. The concept of base rates, or what is called *Bayesian statistics,* is transparently obvious. Suppose that of every thousand men, two hundred are laborers and only one is a teacher. Someone describes to you a person who seems like he should be a teacher and asks you what you think he is. If you remember the base rates of what people are, you should still guess laborer because it is more likely. Here is how the base rate concept works with drug testing.

When all is said and done, the use of illegal drugs in a normal high school is low—1 or 2 percent, if that—despite the wild statistics flying about. Now along comes a drug test that purports to detect drug use 98 percent of the time. Only 2 percent are false positives; that is, the test shows that people used drugs when they didn't in 2 percent of those tested. Forgetting for the moment our politicians' willingness to bust up the psychodynamics of 2 percent of innocent families, let's see how this works. As Wolford puts it,

> The proportion of positive results that are truly false follows a Bayesian rule and depends on three factors: the probability of correctly detecting illegal use (sensitivity),

> the probability of classifying a nonuser as a user (lack of specificity or false positive rate), and the prevalence or base rate of activity. Few if any tests are perfect. Tests fail to detect some users and incorrectly classify some nonusers as users. One standard test to detect cocaine use has a published sensitivity of .98 and a published false positive rate of .02. That sounds good, but think about it for a minute. If 1% of the students in a high school are using cocaine and all of the students were subjected to a drug test, then 67% of all positive results would be false positives; the one true user and the three false positives would yield 4 positive results out of a sample of a hundred. If 5% of the students in a high school are using cocaine, then 28% of all positive results still would be false positives.

So think about it. Under the first scenario the odds are two to one that someone who tests positive does not use drugs. This is exactly the opposite of what people think. Under the second scenario it's still almost one in three. As a practical matter, the tests actually used have much worse sensitivity—and so an even worse predictive value. Just think how many families could become unglued because they are told their darling child is a drug user when in fact she isn't. It is horrendous public policy, based on an untrue sense of what such a test means.

Consider the famous Linda problem, a thought problem made up by and adapted from Kahneman and Tversky: Linda is thirty-one years old, single, outspoken, and very bright. She majored in philosophy. As a student, she was deeply concerned with issues of discrimination and social justice and also participated in antinuclear demonstrations.

Please rank the following statements by their probability: (1) Linda is a bank teller. (2) Linda is a bank teller and is active in the feminist movement. Believe it or not, many people say that the latter option is more probable even though this is impossible because having two conditions versus one condition on an outcome makes it less likely that Linda will be a feminist bank teller.

Examples of this sort of thing abound, all leading to the view that humans are "probability blind," as the Italian cognitive scientist Masimo Piatelli-Palmarini has proposed. Or, as a stormy student said after Steve Pinker's MIT lecture on the topic: "I am ashamed for my species." How can we be so bad? Therein lies a tale, and the moral goes something like this: If you find a sport that Michael Jordan isn't good at, does that mean he is not a great athlete? Critics complain that psychologists' laboratory games create the false impression that we are lousy at reasoning. But the games exemplify when rational processes go awry, just as when our visual system gets tripped up by illusions. The reason this is such a big deal—why there is so much heat on the issue of rationality, or the lack of it—is that from an evolutionary standpoint the wonder of the human brain is its ability to reason sensibly and hence enhance reproductive success.

I don't know Lola Lopes at the University of Iowa, but I would like to. She has written with great brilliance and force about these issues; her conclusion is that characterizing humans as lousy decision makers is wrongheaded. This story goes back to the forties, when mathematicians John von Neuman and Oskar Morgenstern published a now classic book, *Theory of Games and Economic Behavior,* that outlines how we reach decisions. Game theory revolves around the idea of expected utility (EU) of a decision. The book immediately drew mathematically oriented psychologists to the idea that human decisions can be understood through normative models. The concept of "normative" means that people should and can think completely rationally if they want to understand how probable it is that an event may occur. The idea is that not only do humans respond accordingly, but that you can tell them how to respond if all the factors going into a utility judgment are worked out. Psychologist Leonard James Savage extended the idea of EU into something called "subjective" or "personal probabilities" (SEU) and focused experimental psychologists on the issue of how humans decide things. He believed and promulgated the idea that there is a solid mathematics behind the idea that one can predict the probability of a single event occurring. No one really believes this today, but his ideas had a large impact in the 1950s. Lopes recounts how mixed the results were. Some studies supported the idea that humans have wonderful intuitive and statistical horse sense; others pointed out where we fall down. The studies, highly quantitative, never took hold in mainstream psychology—until Tversky and Kahneman presented their vignettes, catchy examples of our shortcomings.

In retrospect maybe the problems were stacked against the subjects. While subjects could solve each problem correctly using probability theory, each was so preciously close to everyday intuitions that the latter mechanism tended to rule. It is as if one were to cook up a bunch of problems and throw them at a crowd. They readily dispense with most of the problems, but find a few tongue-tying; these are the ones we hear about. In many ways this is not unlike my examples of perceptual illusions. Our percepts bounce back and forth because the data are unstable, and so are our decisions.

Lopes recounts a famous example that Kahneman and Tversky presented in 1972. All families with six children were surveyed in a city. In seventy-two families the exact order of births of boys and girls was GBGBBG. What is your estimate of the number of families surveyed in which the exact order of births was BGBBBB?

The first problem that Lopes and others have with this test is that no matter what people do, the test is set up to make a strong inference about the outcome. If you say seventy-two, which is the correct answer, you have a fine sense of probabilities. After all, we know that having either a boy or a girl is a fifty-fifty proposition, so all sequences of equal length and with equally probable outcomes have the same probability. If seventy-two families have a particular sequence of six births, another seventy-two ought to have another equally probable sequence. Bravo for you.

What the problem leads you to think, however, is that they are asking about a *sequence* of births, and so one of your representations springs into action—which is to say that one of our rule-of-thumb heuristics comes into play. The fifty-fifty rule leads you to think the GBGBBG was more representative. And, sure enough, that is what people often guess. The inference therefore is that we are lousy at intuitive statistics and lousy at reasoning.

I have been around academics all my life and have noted one crowning feature in their view of the world. Even though adorned with an armament of years of education, off-the-shelf attitudes about promoting human understanding, and cant about diversity and community, the first thing out of their mouths about almost anyone is that he or she is either stupid or smart. So along comes a couple of smart and engaging scientists like Kahneman and Tversky suggesting that there can be pitfalls in how we think. Suddenly everybody seizes on this as proof that our species is stupid. Rejoice!

Indeed, one review of who cites what in this field found that during the seventies thirty-seven articles showed how good we are as a species and forty-seven reported poor performance. As Lopes summarizes the situtation, "Although articles reporting good performance and articles reporting poor performance were published in comparable numbers and in journals of comparable visibility, reports of poor performance were cited an average of 27.8 times in the sampled period whereas reports of good performance were cited only 4.7 times." Scholars outside psychology seized on the report. About 20 percent of the citations came from outside the field, and of those, 100 percent used the references citing how poor the human is as a decision-making device.

I mentioned before that everyone in the business of minds and brains and evolutionary theory is exercised about all of this, and for good reason. Lyndon Johnson's aphorism, "He may be a dog, but he is my dog and don't you kick him," comes to mind. The human rational system is at stake. Sure you can trick it, but it is a system that brought us a long way, and anyone saying now that humans are a towering inferno of irrationality ought to be challenged, and aggressively so. After all, if we ask what the brain is for, we quickly realize its whole raison d'être is to make decisions—especially concerning our reproductive success. To claim we are lousy decision makers because we can't think clearly is a hard pill to swallow. We may be poor at math or even have limited statistical capacities, but to claim we are irrational is over the top. As Lopes says,

> We cannot overlook the fact that formal probability theory develops very late in human intellectual history. . . . Had probabilistic thinking truly been as central to good action as geometry and arithmetic, we believe it would have arisen sooner and would underlie more of rational thought. . . . Rationality, on the other hand, is so much a human artifact that one never even speaks of artificial rationality. Like all things human, this rationality is limited in its scope and usefulness in a complex world. Intelligence is bigger, murkier, and more serious stuff. It has had to keep us alive as a species and as individuals even at the cost of error. Robustness, generality, and practicability are the demands of life itself. They reside in intelligence because they must.

In many of today's attempts to figure out rational processes, subjects are presented with artificial thought problems developed in the laboratory. But if our brains are adapted only to real-world problems, how applicable are the results? Many logic tests stump students because they are posed in the abstract. Does their failure mean that students are illogical? It doesn't because when the same logic problem is based on a real-life story about a social exchange such as obtaining beer or food, their logic systems work just fine.

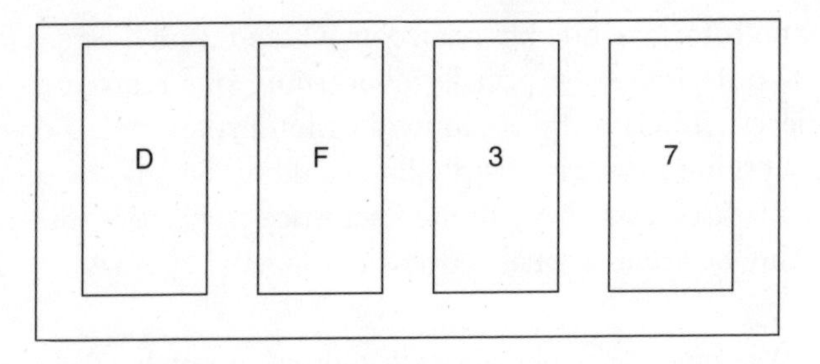

Figure 5

Leda Cosmides at the University of California, Santa Barbara, has worked out a telling example of this fact. She starts by using a test that shows how poorly even educated people can perform a simple logical task. Try it yourself. Each card in a stack has a number on one side and a letter on the other. Examine the four cards in Figure 5. The task is simple: Determine whether the following rule has any exceptions. If a card has a D on one side, it has a 3 on the other side. Which cards do you have to turn over to discover if this is true? Feel your mind rattling? Consider now the following problem: You are a bouncer in a bar, and your job is to make sure that no one under twenty-one drinks beer. The cards displayed have information about age on one side and what the patron is drinking on the other, as in Figure 6. Again, which cards do you need to turn over? The mind springs to action on this task. Obviously, it is the first and last card, as in the more abstract example.

Alas, so much of what has been written about our species being irrational is baloney. As Pinker reminds us, the

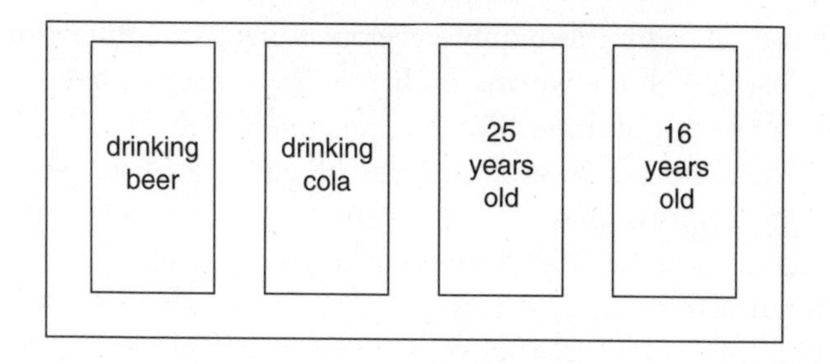

Figure 6

species that invented the concept of probability can't be ignorant of it. Tricks that make our skills seem crude and irrational can be played, but one has to dig deeper and see that the human being usually has the requisite mental skills to make proper inferences. Trouble starts when a problem is misleading or difficult, requiring several steps before a solution is possible. This is when the interpreter works overtime. The natural hypothesis-generating system continues on to suggest why a certain pattern of information is the way it is.

. . .

Imagine a Thing that is full of clever special devices that allow it to reproduce, navigate around its environment, detect unwanted predators, and perform a multitude of tasks. Then imagine the Thing with a new feature. The Thing asks "how?" How does X relate to Y? Just imagine the consequences of having this added feature. Although we know a lot about this Thing, the brain, we struggle to figure out why this highly developed piece of equipment makes us feel like we are in charge. We don't feel like we are a bag full of tricks. We feel integrated. Why?

The brain is built so the genetic blueprint for a species is in tight control of development. As neurons are made and migrate to their final destination, they do so in a physiochemical milieu. The neurons start functioning and become physiologically active, firing off and processing information. Both their physiochemical environment and their physiological activity play a role in the final disposition of the neuronal network. But the genetic system has already taken that into account. Those environmental forces are rather constant, so the genetic coding required to find neurons making their way to targets is established. As I mentioned earlier, the manipulation of one gene during development can disrupt one layer of cells in one area of the brain. That couldn't be done if a neural free-for-all were going on during development. The brain gets built, and it gets built from a blueprint.

As soon as the brain is built, it starts to express what it knows, what it comes with from the factory. And the brain comes loaded. The number of special devices that are in place and active is staggering. Everything from perceptual phenomena to intuitive physics to social exchange rules comes with the brain. These things are not learned; they are innately structured. Each device solves a different problem. Not to recognize this simple truth is to live in a dream world.

When animals' fixed behaviors are revealed as automatic and built in, no one blinks at that. People get nervous, though, when the same sort of arrangement is suggested for human perceptual and cognitive functions. But there is no need to fret: A moment's thought should tell you that your capacities for rational thought and navigating around in the world are built-in features of your brain. If each brain had to learn all the rules of just one complex mental skill, people would be wandering around aimlessly in the dark.

The multitude of devices we have for doing what we do are factory installed; by the time we know about an action, the devices have already performed it. Brain imaging techniques allow us to see how and where the brain is active before a behavior is actually executed. The decision has already been made when our conscious self catches up with these activities and declares we have made the decision.

There is no accounting for taste, but surely one of life's little pleasures, or horrors, is seeing how an idea, an impulse, plays out on the printed page. As a sentence or paragraph travels from the fingertips to the computer screen, we marvel at the way things get said. Of course we think all the verbal associations careening through our lexicon, generating preferred word sequences and unexpected colorings of our thoughts, reflect our conscious mind. Yet there is no way on earth we could actively be structuring the fine details of a statement. We are adjusting, perhaps, the possibilities being served up, but brain processes outside of our awareness are doing the serving.

You know how active the automatic brain is, especially when you are awakened at 3 A.M. by a profusion of concerns swirling around in your mind. You can't get back to sleep because your brain is in charge. You observe the automatic brain functioning in social settings when an attractive member of our species passes by your gaze. You see it at work as your mood switches unexpectedly. You marvel at how quickly you can flare up when something challenges you. From visual illusions, to grasping onto a tree branch, to carrying out syllogisms, your brain is working for you. How could it be any other way?

"Goddamn it, I am me and I am in control." Whatever it is that brain and mind scientists are finding out, there is no way they can take that feeling away from each and every one of us. Sure, life is a fiction, but it's our fiction and it feels good and we are in charge of it. That is the sentiment we all feel as we listen to tales of the automatic brain. We don't feel like zombies; we feel like in-charge, conscious entities—period.

This is the puzzle that brain scientists want to solve. Prudent people bug out at this point and just leave the brain facts on the table. But I do have a thought about the gap between our understanding of the brain and the sensation of our conscious lives. There is a deep belief that we can attain not only a neuroscience of consciousness, but also a neuroscience of *human* consciousness. It is as if something terribly new and complex happens as the brain enlarges to its human form. Whatever this is, it triggers our capacity for self-reflection, for ennui, for lingering moments.

I propose a three-step process for knowing how the brain enables conscious experience. First, we should focus on what we mean when we talk about conscious experience; it is merely our awareness we have of our capacities as a species, not the capacities themselves—only the awareness or feelings we have about them. The brain is not a general-purpose computing device; it is a collection of circuits devoted to specific capacities. Though this is true for all brains, the wonderful thing about the human brain is that we have untold numbers of these capacities. We have more than the chimp, which has more than the monkey, which has more than the cat, which runs circles around the rat. So first we must distinguish a species' capacities from its feelings about those capacities.

Second, each species is aware of its capacities. Can there be any doubt that a rat at the moment of copulation is as sensorially fulfilled as a human? Of course it is. Do you think a cat doesn't enjoy a good piece of cod? Of course it does. Or a monkey doesn't enjoy a spectacular swing? Again, it has to be true. So what is human consciousness? It is the very same awareness, save for the fact that we can be aware of so much more, so many things. A circuit—perhaps a single system or one duplicated over and over again—is associated with each brain capacity. The more systems a brain possesses, the greater its awareness of capacities.

Think of the variations in capacity within our own species; they are not unlike the vast differences between species. Years of split-brain research informs us that the left hemisphere has many more mental capacities than the right. The right hemisphere's level of awareness is limited; it knows precious little about a lot of things. But limits to human capacity are everywhere in the population. No one need be offended to realize that just as one person with normal intelligence can understand Ohm's law, others, like yours truly, are clueless about quantum mechanics. I am unable to be aware of what quantum theory means for the universe. The circuits that enable me to understand these things are not present in my brain.

When focusing on circuits that arise from natural selection, we realize that the brain is not a unified neural network that supports general problem solving. If we accept this realization as fact, we can concentrate on the possibility that smaller, more manageable circuits create awareness of a species' capacities. Holding fast to the notion of a unified neural network means we can understand conscious experience only by figuring out the interactions of billions of neurons. That task is hopeless. My scheme is not.

The very same split-brain research that exposed shocking differences between the two hemispheres also revealed that the left hemisphere contains the interpreter, whose job is to interpret our behavior and our responses, whether cognitive or emotional, to environmental challenges. The interpreter constantly establishes a running narrative of our actions, emotions, thoughts, and dreams. It is the glue that unifies our story and creates our sense of being a whole, rational agent. It brings to our bag of individual instincts the illusion that we are something other than what we are. It builds our theories about our own life, and these narratives of our past behavior pervade our awareness.

So the problem of consciousness is tractable. We don't have to find the code of one huge interacting neural network. The third step is to find the common and perhaps simple neural circuit(s) that allows vertebrates to be aware of their species-specific capacities. The enabling circuit(s) in the rat is most likely present in the human brain. Understanding this makes the problem scientifically resolvable. What makes us so grand is that the circuit has much more to work with in the human brain.

Finally things become clear. The insertion of an interpreter into an otherwise functioning brain delivers all kinds of by-products. A device that asks how infinite numbers of things relate to each other and gleans productive answers to that question can't help but give birth to the concept of self. Surely one of the questions the device would ask is "Who is solving these problems?" Call that "me," and away the problem goes! The device that has rules for solving a problem of how one thing relates to another must be reinforced for such an action, just as an ant's solving where the daily meal is reinforces its food-seeking devices.

Our brains are automatic because physical tissue carries out what we do. How could it be any other way? The brain does it before our conceptual self knows about it. But the conceptual self grows and grows and reaches proportions where the biological fact makes an impact on our consciousness but doesn't paralyze us. The interpretation of things past liberates us from the sense of being tied to the demands of the environment and produces the wonderful sensation that our self is in charge of our destiny. All our everyday success at reasoning through life's data convinces us of our centrality. Because of that, maybe we can drive our automatic brains to greater accomplishments and enjoyment of life.

BIBLIOGRAPHY

Chapter 1. The Fictional Self

Chomsky, N. Knowledge of language: Its elements and origins. *Philosophical Transactions of the Royal Society of London* 295 (1981): 223–234.

Dawkins, R. *Climbing Mount Improbable*. New York: W. W. Norton, 1996.

Gazzaniga, M. S. *Nature's Mind*. New York: Basic Books, 1992.

Goodall, J. *The Chimpanzees of Gombe*. Cambridge: Harvard University Press, 1986.

Gould, Stephen J. Exaptation: A crucial tool for an evolutionary psychology. *Journal of Social Issues* 47 (1991): 43–65.

James, W. *The Principles of Psychology*, vol. I. New York: Dover, 1890.

Jerne, N. Antibodies and learning: Selection versus instruction. In *The Neurosciences: A Study Program*, vol. 1 (G. Quarton, T. Melnechuck, and

F. O. Schmidt, eds.). New York: Rockefeller University Press, 1968.

Miller, G. A. Interview. In *Fundamentals of Cognitive Neuroscience* (M. S. Gazzaniga, R. B. Ivry, and G. R. Mangun, eds.). New York: W. W. Norton, 1998.

Nesse, R. M., and Williams, G. C. *Why We Get Sick?* New York: Vintage Books, 1996.

Pinker, S. *The Language Instinct.* New York: W. W. Norton, 1994.

Pinker, S. *How the Mind Works.* New York: W. W. Norton, 1997.

Premack, D., and Premack, A. J. Why animals have neither culture nor history. In *Companion Encyclopedia of Anthropology.* London: Routledge, 1994, pp. 350–365.

Quartz, S. R., and Sejnowski, T. J. Controversies and issues in developmental theories of mind: Some constructive remarks. In *Behavioral and Brain Sciences.* New York: Cambridge University Press (in press).

Rozin, P., and Schull, J. The adaptive-evolutionary point of view in experimental psychology. In *Steven's Handbook of Experimental Psychology.* New York: Wiley, 1988.

Updike, J. *Self Consciousness.* New York: Knopf, 1989.

Trivers, R. L. The evolution of reciprocal altruism. *Quarterly Review of Biology* 46 (1971): 35–57.

Weiskrantz, L. *Blindsight: A Case Study and Implications.* New York: Oxford University Press, 1986.

Wilson, E. O. *Sociobiology: The New Synthesis.* Cambridge: Harvard University Press, 1974.

Chapter 2. Brain Construction

Gallistel, C. R. The replacement of general-purpose theories with adaptive specializations. In *The Cognitive Neuroscience* (M. S. Gazzaniga, ed.). Cambridge: MIT Press, 1995, pp. 1255–1267.

Goodman, C. S., and Shatz, C. J. Developmental mechanisms that generate precise patterns of neuronal connectivity. *Cell* 72 (suppl) (1993): 77–98.

Herschkowitz, N., Kagan, J., and Zilles, K. Brain bases for behavioral development in the first year. *Pediatric Neurology* (in press).

Hubel, D. H., and Weisel, T. N. Receptive fields, binocular interaction and functional architecture in the cat's visual cortex. *Journal of Physiology* 160 (1962): 106–154.

Huntley G. W. Correlation between patterns of horizontal connectivity and the extent of short-term representational plasticity in rat motor cortex. *Cerebral Cortex* 7(2) (1997): 143–156.

Innocenti, G. M. Exuberant development of connections, and its possible permissive role in cortical evolution. *Trends in Neuroscience* 18 (1995): 397–402.

Jones, E. G. Santiago Ramon y Cajal and the Croonian lecture, March 1894. *Trends in Neuroscience* 17 (1994): 190–192.

Kagan, J. Temperament and the reactions to unfamiliarity. *Child Development* 68 (1997): 139–143.

Kuhl, Patricia K. Infant speech perception: A window on psycholinguistic development. *International Journal of Psycholinguistics* 9 (1993): 33–36.

Marin-Padilla, M. Prenatal development of human cerebral cortex: An overview. *International Pediatrics* 10 (1995): 6–15.

Medin, D. L., and Davis, R. T. Memory. In *Behavior of Nonhuman Primates: Modern Research Trends,* vol. 5 (A. M. Schrien and F. Stollivitz, eds.). New York: Academic Press, 1974, pp. 1–47.

Merzenich, M. M., Grajski, K. A., Jenkins, W. M., Recanzone, G. H., and Peterson, B. Functional cortical plasticity. Cortical network origins of representations changes. *Cold Spring Harbor Symposium on Quantitative Biology* 55 (1991): 873–887.

Merzenich, M. M., Nelson, R. H., Kaas, J. H., Stryker, M. P.,

Jenkins, W. M., Zook, J. M., Cynader, M. S., and Schoppmann, A. Variability in hand surface representations in areas 3b and 1 in adult owl and squirrel monkeys. *Journal of Comparative Neurology* 258 (1987): 281–296.

Merzenich, M. M., Nelson, R. H., Stryker, M. P., Cynader, M. S., Schoppmann, A., and Zook, J. M. Somatosensory cortical map changes following digital amputation in adult monkey. *Journal of Comparative Neurology* 224 (1984): 591–605.

Rakic, P. A small step for the cell, a giant leap for mankind: A hypothesis of neocortical expansion during evolution. *Trends in Neuroscience* 18 (1995): 383–388.

Ramachandran, V. S., Rogers-Ramachandran, D., and Stewart, M. Perceptual correlates of massive cortical reorganization. *Science* 258 (1992): 1159–1160.

Recanzone, G. H., Schreiner, C. E., and Merzenich, M. M. Plasticity in the frequency representation of primary auditory cortex following discrimination training in adult owl monkeys. *Journal of Neuroscience* 13 (1993): 87–103

Shatz, C. J. The developing brain. *Scientific American* 267(3) (1992): 61–67.

Singer, W. Development and plasticity of cortical processing architectures. *Science* 270 (1995): 758–764.

Smith, B. C. The owl and the electric encyclopedia. *Artificial Intelligence* 47 (1991): 251–288.

Sperry, R. W. Mechanisms of neural maturation. In *Handbook of Experimental Psychology* (S. Stevens, ed.). New York: Wiley, 1963, pp. 236–280.

Voyvodic, J. T. Cell death in cortical development: How much? Why? So what? *Neuron* 16 (1996): 693–696.

Wehner, R., and Srinvasan, M. V. Searching behavior of desert ants, genus, Catagyhlphis (Formicidae, Humenoptera). *Journal of Comparative Physiology* 142 (1981): 315–338.

Weliky, M., and Katz, L. C. Disruption of orientation tuning in visual cortex by artificially correlated neuronal activity. *Nature* 17 (1997):386, 680–685.

Chapter 3. The Brain Knows Before You Do

Deecke, L., and Kornhuber, H. H. An electrical sign of participation of the mesial "supplementary" motor cortex in human voluntary finger movement. *Brain Research* 159 (1978):473–476.

Dehaene, S. *The Number Sense.* New York: Oxford University Press, 1997.

Dehaene, S., and Cohen, L. Cerebral pathways for calculation: Double dissociations between Gerstmann's acalculia and subcortical acalculia. *Cortex* 33 (1997):219–250.

Fendrich, R., Wessinger, C. M., and Gazzaniga, M. S. Residual vision in a scotoma: Implications for blindsight. *Science* 258 (1992):1489–1491.

Gaulin, S. J. C, and Fitzgerald, R. W. Sexual selection for spatial-learning ability. *Animal Behavior* 37 (1989):332–331.

Gazzaniga, M. S., Fendrich, R., and Wessinger, C. M. Blindsight reconsidered. *Current Directions in Psychological Science* 3 (1994):93–96.

Gelman, R. Domain specificity in cognitive development: Universals and non-universals. *International Journal of Psychology* (in press).

Holtzman, J. D. Interactions between cortical and subcortical visual areas: Evidence from human commissurotomy patients. *Vision Research* 24 (1984):801–813.

Holtzman, J. D., Sidtis, J. J., Volpe, B. T., Wilson, D. H., and Gazzaniga, M. S. Dissociation of spatial information for stimulus localization and the control of attention. *Brain* 104 (1981):861–872.

Libet, B. Neural processes in the production of conscious experience. *Brain* 172 (1979): 96–110.

Libet, B. Neural time factors in conscious and unconscious mental functions. In *Toward a Science of Consciousness: The First Tucson Discussions and Debates* (S. R. Hameroff et al., eds.). Cambridge: MIT Press, 1996, pp. 337–347.

Libet, B., Pearl, D. K., Morledge, D. E., Gleason, C. A., Hosobuchi, Y., and Barbaro, N. M. Control of the transition from sensory detection to sensory awareness in man by the duration of a thalamic stimulus. *Brain* 114 (1991): 1731–1757.

Libet, B., Wright, E. W., and Gleason, C. A. Readiness-potentials preceding unrestricted "spontaneous" vs. pre-planned voluntary acts. *Electroencephalography and Clinical Neurophysiology* 54 (1982): 322–335.

Nijhawan, R. Visual decomposition of color through motion extrapolation. *Nature* 386 (1997): 66–69.

Sidtis, J. J., Volpe, B. T., Holtzman, J. D., Wilson, D. H., and Gazzaniga, M. S. Cognitive interaction after staged callosal section: Evidence for a transfer of semantic activation. *Science* 212 (1981): 344–346.

Volpe, B. T., LeDoux, J. E., and Gazzaniga, M. S. Information processing of visual stimuli in an extinguished field. *Nature* 282 (1979): 722–724.

Wessinger, C. M., Fendrich, R., and Gazzaniga, M. S. Islands of residual vision in hemianopic patients. *Journal of Cognitive Neuroscience* 9 (1997): 203–222.

Chapter 4: Seeing Is Believing

Cavanagh, P. Attention-based motion perception. *Science* 257 (1992): 1563–1565.

Cavanagh, P. Predicting the present. *Nature* 386 (1997): 19–21.

Crick, F., and Koch, C. Are we aware of neural activity in primary visual cortex? *Nature* 375 (1995): 121–123.

Friedman-Hill, S. R., Robertson, L. C., and Treisman, A. Parietal contributions to visual feature binding: Evidence from a patient with bilateral lesions. *Science* 269 (1995): 853–855.

Gibson, J. J. *The Ecological Approach to Visual Perception*. Boston: Houghton Mifflin, 1979.

He, S., Cavanagh, P., and Intriligator, J. Attentional resolution. *Trends in Cognitive Sciences* (1997): 115–121.

Nakayama, K. Gibson—An appreciation. *Psychological Review* 101 (1994): 329–335.

Shepard, R. N. Psychological relations and psychophysical scales: On the status of "direct" psychophysical measurement. *Journal of Mathematical Psychology* 24 (1981): 21–57.

Shepard R. N. *Mind Sights*. San Francisco: W. H. Freeman, 1990.

Shepard, R. N. Perceptual-cognitive universals as reflections of the world. *Psychonomic Bulletin and Review* 1 (1994): 2–28.

Treisman, A. Search, similarity, and integration of features between and within dimensions. *Journal of Experimental Psychology: Human Perception and Performance* 17 (1991): 652–676.

Treisman, A. Perceiving and re-perceiving objects. *American Psychologist* 47 (1992): 862–875.

Treisman, A. The perception of features and objects. In *Attention: Selection, Awareness, and Control: A Tribute to Donald Broadbent* (A. Baddley and L. Weiskrantz, eds.). Oxford: Clarendon Press, 1993, pp. 5–35.

Treisman, A. and Sato, S. Conjunction search revisited. *Journal of Experimental Psychology: Human Perception and Performance* 16 (1990): 459–478.

Chapter 5: The Shadow Knows

Arad, Z., Midtgård, U., and Bernstein, M. Thermoregulation in turkey vultures: Vascular anatomy, anteriovenous heat exchange, and behavior. *The Condor* 91(3) (1989): 505–514.

Bechera, A., Damasio, H., Tranel, D., and Damasio, A. Deciding advantageously before knowing the advantageous strategy. *Science* 275 (1997): 1293–1294.

Goodale, M. A., and Milner, A. D. Separate visual pathways for perception and action. *Trends in Neurosciences* 15 (1992): 20–25.

Haffenden, A., and Goodale, M. A. The effect of pictorial illusion on prehension and perception. *Journal of Cognitive Neuroscience* 10 (1998): 122–136.

Leopold, D. A., and Logothetis, N. K. Activity changes in early visual cortex reflect monkeys' percepts during binocular rivalry. *Nature* 379 (1996): 549–553.

Logothetis, N. K., Leopold, D. A., and Sheinberg, D. L. What is rivalling during binocular rivalry? *Nature* 380 (1996): 621–624.

Logothetis, N. K., Pauls, J., and Poggio, T. Shape representation in the inferior temporal cortex of monkeys. *Current Biology* 5 (1995): 552–563.

Marchand, J.-C. *La maladie dans la vie d'Emmanuel Kant: Diagnostic rétrospectif et causalité neuropsychologique*. Thèse de doctorat. Université Paris VII, 1996.

Palmer, R. S. (ed.). *Handbook of North American Birds*, vol. 4. New Haven: Yale University Press, 1988.

Panum, P. L. *Physiologische Untersuchungen über das Sehen mit zwei Augen*. Kiel: Schwers, 1858.

Platt, M., and Glimcher, P. Representation of prior probability and movement metrics by area LIP neurons. *Society of Neuroscience Abstract,* 1997.

Poole, A. F. *Ospreys: A Natural and Unnatural History*. Cam-

bridge, England: Cambridge University Press, 1989.
Servos, P., and Goodale, M. Preserved visual imagery in visual form agnosia. *Neuropsychologia* 33 (1995): 1383–1394.
Servos, P., Martin, L., and Goodale, M. A. Dissociation between two modes of spatial processing by a visual form agnostic. *NeuroReport* 6 (1996): 1893–1896.
Shadlen, M. N., Britten, K. H., Newsome, W. T., and Movshon, J. A. A computational analysis of the relationship between neuronal and behavioral responses to visual motion. *Journal of Neuroscience* 16(4) (1996): 1486–1510.
Sheinberg, D. L., and Logothetis, N. K. The role of temporal cortical areas in perceptual organization. *Proceedings of the National Academy of Sciences of the United States of America* 94 (1997): 3408–3413.
Smith, S. A., and Paselk, R. A. Olfactory sensitivity of the turkey vulture (*Cathartes aura*) to three carrion-associated odorants. *The Auk* 103 (1986): 586–592.

Chapter 6. Real Memories, Phony Memories

Baynes K., Wessinger, C. M., Fendrich, R., and Gazzaniga, M. S. The emergence of the capacity of the disconnected right hemisphere to control naming: Implications for functional plasticity. *Neuropsychologia* 33 (1995): 1225–1242.
Ceci, S. J., Huffman, M. L., and Smith, E. Repeatedly thinking about a non-event: Source misattributions among preschoolers. *Consciousness and Cognition* 3 (1994): 388–407.
Ceci, S. J., and Loftus, E. F. "Memory work": A royal road to false memories? *Applied Cognitive Psychology* 8 (1994): 351–364.
Chen, C., and Tonegawa, S. Molecular genetic analysis of synaptic plasticity, activity-dependent neural development, learning, and memory in the mammalian brain. *Annual Review of Neuroscience* 20 (1997): 157–184.

Festinger, L. *A Theory of Cognitive Dissonance*. Evanston, IL: Row, Peterson, 1957.

Gazzaniga, M. S. *The Social Brain*. New York: Basic Books, 1985.

Gazzaniga, M. S., Eliassen, J. C., Nisenson, L., Wessinger, C. M., and Baynes, K. B. Collaboration between the hemispheres of a callosotomy patient—Emerging right hemisphere speech and the left brain interpreter. *Brain* 119 (1996): 1255–1262.

Gazzaniga, M. S., and LeDoux, J. E. *The Integrated Mind*. New York: Plenum Press, 1978.

Metcalfe, J., Funnell, M., and Gazzaniga, M. S. Right hemisphere superiority: Studies of a split-brain patient. *Psychological Science* 6 (1995): 157–164.

Miller, M. B., and Gazzaniga, M. S. Creating false memories for visual scenes. *Neuropsychologia* (in press).

Penfield, W., and Roberts, L. *Speech and Brain Mechanisms*. Princeton, NJ: Princeton University Press, 1959.

Phelps, E. A., and Gazzaniga, M. S. Hemispheric differences in mnemonic processing: The effects of left hemisphere interpretation. *Neuropsychologia* 30 (1992): 293–297.

Roediger, H. L. III, and McDermott, K. B. Creating false memories: Remembering words not presented in lists. *Journal of Experimental Psychology: Learning, Memory, and Cognition* 21 (1995): 803–814.

Schacter, D. L., Norman, K. A., and Koutstaal, W. The cognitive neuroscience of constructive memory. *Annual Review of Psychology* (in press).

Schindler, R. A., and Kessler, D. K. The UCSF/Storz cochlear implant: Patient performance. *American Journal of Otology* 8 (1987): 247–255.

Simons, D. J. In sight, out of mind: When object representations fail. *Psychological Science* 7 (1996): 301–305.

Sperry, R. W., Gazzaniga, M. S., and Bogen, J. E. Interhemispheric relationships: The neocortical commissures, syndromes of hemisphere disconnection. In *Handbook of Clinical Neurology*, vol. 4 (P. J. Vinken and G. W. Bruyn, eds.). Amsterdam: North-Holland, and New York: Wiley, 1969, pp. 273–290.

Chapter 7. The Value of Interpreting the Past

Barkow, J. H., Cosmides, L., and Tooby, J. *The Adapted Mind: Evolutionary Psychology and the Generation of Culture*. New York: Oxford University Press, 1992.

Festinger, L. *The Human Legacy*. New York: Columbia University Press, 1983.

Gazzaniga, M. S., and Smylie, C. S. Dissociation of language and cognition: A psychological profile of two disconnected right hemispheres. *Brain* 107 (1983): 145–153.

Goel, V., Gold, B., Kapur, S., and Houle, S. Neuroanatomical correlates of human reasoning. *Journal of Cognitive Neuroscience* (in press).

Kahneman, D. Judgment and decision making: A personal view. *Psychological Science* 2 (1991): 142–145.

Kahneman, D., Slovic, P., and Tversky, A. (eds). *Judgment Under Uncertainty: Heuristics and Biases*. Cambridge, England: Cambridge University Press, 1982.

Kahneman, D., and Tversky, A. Subjective probability: A judgment of representativeness. *Cognitive Psychology* 3 (1972): 430–454.

Kahneman, D., and Tversky, A. Can irrationality be intelligently discussed? *Behavioral and Brain Sciences* 3 (1983): 509–510.

Kahneman, D., and Varey, C. A. Propensities and counterfactuals: The loser that almost won. *Journal of Personality and Social Psychology* 59 (1990): 1101–1110.

Lopes, L. L. The rhetoric of irrationality. *Theory and Psychology* 1 (1991):65–82.

Lopes, L. L., and Oden, G. C. The rationality of intelligence. In *Rationality and Reasoning* (E. Elles and T. Maruszewski, eds.). Amersterdam: Rodopi, 1987.

Piatelli-Palmarini, M. *Inevitable Illusions*. New York: Wiley, 1994.

Pinker, S. *How the Mind Works*. New York: W. W. Norton, 1997.

Savage L. J. The sure-thing principle. In *The Foundations of Statistics*. New York: Wiley, 1954.

Simon, H. A. Rational choice and the structure of the environment. *Psychological Review* 63 (1956):129–138.

von Neuman, J., and Morgenstern, O. *Theory of Games and Economic Behavior*. Princeton, NJ: Princeton University Press, 1944.

Wolford, G. *False Positives Can Kill You*. Berkeley: University of California Press (in press).

INDEX

Credits

Figure 1, the turning tables illusion, is from *Mind Sights* by Shepard, © 1990 by Roger N. Shepard. Used with permission of W. H. Freeman and Company.

Figure 2, the limits of attention, is adapted from S. He, P. Cavanagh, and J. Intriligator, Attentional resolution, *Trends in Cognitive Sciences* (1997): 115–121.

Figure 3, the Ebbinghaus illusion, is adapted from A. Haffenden and M. A. Goodale, The effect of pictorial illusion on prehension and perception, *Journal of Cognitive Neuroscience* 10 (1998): 122–136.

Also by Michael S. Gazzaniga

GENERAL READING

The Social Brain

Mind Matters

Nature's Mind

Conversations in Cognitive Neuroscience

RESEARCH MONOGRAPHS

The Bisected Brain

The Integrated Mind (with Joseph LeDoux)

TEXTBOOKS

Fundamentals of Psychology

Functional Neuroscience (with Diana Steen and B. T. Volpe)

Psychology

Fundamentals of Cognitive Neuroscience (with R. B. Ivry and G. R. Mangun)

Good Readings in Psychology (with E. P. Lovejoy)

EDITED SCHOLARLY WORKS

Handbook of Psychobiology (with Colin Blakemore)

Handbook of Cognitive Neuroscience

Perspectives in Memory Research

Neuropsychology: Handbook of Behavioral Neurobiology (vol. 2)

Extending Psychological Frontiers: Selected Works of Leon Festinger

(with Stanley Schachter)

The Cognitive Neurosciences

Designer: Steve Renick
In-text figures: Bill Nelson
Chapter opening art and
brain figure (p. 117): M. Alex Meredith
Compositor: G&S Typesetters, Inc.
Text: 11/13 Bembo
Display: Gill Sans
Printer: Haddon Craftsmen, Inc.
Binder: Haddon Craftsmen, Inc.